Molecular Genetics
and Comparative Evolution

RESEARCH STUDIES IN BOTANY AND RELATED APPLIED FIELDS

Series Editor: **Dr. P. S. Nutman, F.R.S.**

* Out of print

Molecular Genetics and Comparative Evolution

J. Langridge
Commonwealth Scientific and Industrial
Research Organisation,
Canberra, Australia

RESEARCH STUDIES PRESS LTD.
Taunton, Somerset, England

JOHN WILEY & SONS INC.
New York · Chichester · Toronto · Brisbane · Singapore

RESEARCH STUDIES PRESS LTD.
24 Belvedere Road, Taunton, Somerset, England TA1 1HD

Marketing and Distribution:

Australia and New Zealand:
JACARANDA WILEY LTD.
GPO Box 859, Brisbane, Queensland 4001, Australia

Canada:
JOHN WILEY & SONS CANADA LIMITED
22 Worcester Road, Rexdale, Ontario, Canada

Europe, Africa, Middle East and Japan:
JOHN WILEY & SONS LIMITED
Baffins Lane, Chichester, West Sussex, England

North and South America:
JOHN WILEY & SONS INC.
605 Third Avenue, New York, NY 10158, USA

South East Asia:
JOHN WILEY & SONS (SEA) PTE LTD
37 Jalan Pemimpin #05-04
Block B Union Industrial Building, Singapore 2057

Library of Congress Cataloging in Publication Data

Langridge, J. (John), 1923–
    Molecular genetics and comparative evolution / J. Langridge.
        p.   cm. — (Research studies in botany and related applied
fields ; 9)
    Includes bibliographical references and index.
    ISBN 0 86380 106 4. — ISBN 0 471 92876 3 (Wiley)
    1. Evolution.   2. Molecular genetics.   I. Title.   II. Series.
QH371.L33   1991
575--dc20   90-8986
                                                                                  CIP

British Library Cataloguing in Publication Data

Langridge, J.
    Molecular genetics and comparative evolution.
    1. Organisms. Evolution. Genetic factors.
    I. Title.   II. Series.
    575.1

    ISBN 0 86380 106 4

    ISBN 0 86380 106 4 (Research Studies Press Ltd.)
    ISBN 0 471 92876 3 (John Wiley & Sons Inc.)

Typeset by B.A.T.S. Bureau, Taunton, England
Printed in Great Britain by SRP Ltd., Exeter

PREFACE

An understanding of evolution, by which is meant the gradual development of complex organisms from their primeval beginnings, has been gradually advancing only from the early nineteenth century. Before that time, the doctrine of special creation, based on the first two chapters of the book of Genesis, held that the world and all the organisms in it were created by God in 4305 BC. Not until 1801 when Lamarck proposed an inheritance of acquired characters was 'descent with modification' commonly accepted. But the full mechanism remained unknown even though Darwin and Wallace provided evidence in 1858 that evolution occurs by the natural selection of inherited variations. In the modern theory of neo-Darwinism, natural selection is combined with Mendelian genetics with emphasis on the random nature of mutation and the importance of isolating mechanisms. However, the theory is still incomplete in that it does not account for the frequently episodic nature of phylogenetic change and the prolonged evolutionary stasis of many groups, and in that it is unable to predict future evolutionary trends.

In this book we will attempt to contribute to a more complete theory by combining molecular, genetical and evolutionary data and reconsidering the evolutionary significance of genetic variation. Accordingly, this book stresses, not unity of process, but evolutionary differences as expressed in the various living phyla and classes. If we examine the variety of mechanisms generating change in organisms, their distinctive levels of complexity and integration, the adaptive nature of their existences and other matters related to the ease and direction of evolution, certain deviations from accepted evolutionary principles become apparent. They seem to show that, depending on the group under examination, evolution may involve different mechanisms of genetic change, take a variety of courses and reach quite diverse end-points.

These conclusions also take into account molecular data on the origins and relationships of genes. However, the molecular information that bears on the mechanisms of evolutionary change is still small in amount and tentative in significance. I hope that the book will be capable of withstanding the expansion or modification, and even the removal of pieces of evidence shown to be false, that future research will undoubtedly necessitate. So, in the writing, I have tried to keep in mind the caveat of the protozoologist Minchin (1912): 'The views put forward by any man but represent the visions conjured up by his imagination, more or less limited, or intuition, more or less fallacious, of an infinite world of natural phenomena'.

I should like to thank J.H. Campbell of the University of California, M.R. Walter of the Bureau of Mineral Resources, A.J. Gibbs and G.L. Miklos of the Research School of Biological Sciences and R.D. Brock and A.H.D. Brown of the Division of Plant Industry for their opinions on various chapters. I am grateful to my wife and to Mrs P. French and her staff for the careful typing of the manuscript.

Canberra 1990   John Langridge

CONTENTS

INTRODUCTION

Strictly speaking, the secure data identifying the fact and course of evolution come only from fossil remains of organisms from the remote past. In this sense, its data base is like that of stratigraphical geology, although not as complete. And all that the fossil record tells us is what has happened; it says nothing about how it happened. So the reason for evolutionary change must be derived from other data which are quite independent of its physical evidence. These data are also, of necessity, observational, for the extreme slowness of evolution means that it cannot be experimentally produced. Thus it has been observed that there is a continual generation of genetic change by processes assumed to be much the same in all organisms. A second observation is the decisive role of natural selection, which is assumed to be all-pervasive. The dynamic component of modern evolutionary theory is then provided by the principles of population genetics as applied to these processes of mutation and selection.

From the above premises, a coherent theory of the evolutionary process is expressed as a set of governing principles. These are: that spontaneous gene changes (mutations) provide genetic variations in gene expression (phenotype) which are than subjected to positive or negative selection by factors of the environment - genetic changes incorporated in the population are thus usually adaptive ones that enhance the survival or reproduction of organisms; further, that evolutionary change is most often by the gradual accumulation of successive point mutations, due to one or a few alterations in the nucleotides of the genetic material, that result in the appearance of a new or modified structure or function. With a gradualistic course of evolution thus envisaged, the transformation from one adaptive category, usually a population of interbreeding organisms, to another represented by a species, is an unbroken process with further changes of precisely the same kind leading in turn to higher taxonomic units.

The species is taken to be the basic evolutionary unit and evolutionary change from one taxonomic category or taxon to another progresses in the direction: species, genus, family and order.

These principles have been deduced, for the most part, from the study of evolutionary phenomena in the animal kingdom, which has been justified on the grounds that only the vertebrates and the shelled invertebrates leave reliable fossil evidence of the course and direction of evolutionary change. The other four kingdoms of organisms, the prokaryotes, protista, fungi and plants, have been little considered except where their evolution fits the animal pattern. The few treatises on bacterial or plant evolution owe much of their interpretation to the extrapolation of principles derived from the animals.

We may assume that all forms of life started from a single origin about 4000 million years ago and thereafter they became diversified by their differential acquisition of novel forms of organisation. These new organisational features include organelles, multicellularity, cellular differentiation, morphogenesis and nervous systems. Only the more advanced animals have acquired all these new features, which make the nature and extent of their evolution different from that in other organisms. Evolution in animals is mostly based on changes in developmental processes which are much simpler in fungi and plants and practically absent in prokaryotes. These simpler organisms lack the constraints to changes in gene expression that are imposed by the sequential and interactive nature of the development of animal embryos. A further difference is that some adaptations in higher vertebrates may not have genetic causes but be reflections of learning activities which, in turn, are affected by experience. Indeed, it is possible that, in many higher animals, genetic change gives only a rough fit to the environment, with close fitting being provided by behavioural modification.

A genetic advance that has affected the neo-Darwinian view is the use of molecular data to provide a new means of establishing evolutionary mechanisms and relationships independent of fossil evidence. The mechanisms responsible for genetic diversity are various and probably differently distributed in diverse groups. For example, it now appears, at least in some groups, that evolutionary novelty is generated by different types of recombinational variation and not by point mutation.

The above considerations, among others, suggest that it may be informative to take a comparative approach by subdividing evolution according to whether it is rapid, slow or static in a particular group, what the focus of change may be and what mechanisms are involved. Little that is useful can be added to other current evolutionary controversies about continuity or periodicity, adaptation or drift and the role of sexuality, so they will not be topics of this book.

In the first part of the book we consider the various phases through which evolution has gone. Most organisms comprise a number of distinct features such as metabolism, subcellular structure, genome organisation, development and nervous system, so evolution may be subdivided into types depending on which feature undergoes adaptive changes. All features, of course, will be subject to mutational alteration, but changes in some of them have little phenotypic expression, in others they will be mostly deleterious, and only in some will viable changes be selected in the appropriate environment.

Certain of these features, like metabolic pathways and intracellular structure, now undergo few changes that are likely to be adaptive. Their early evolution has provided a satisfactory basis for all forms of life but they are so basic to the organism that they must now remain static or fixed within narrow tolerances. Others, however, like development and, in animals, behaviour, may still vary in significant ways, especially through changes in hormonal activity in development, and learning processes in behaviour. Whereas the basic frameworks of these features may alter very infrequently, modifications of their expression are possible by genetic alterations in regulation.

A survey of types of evolution indicates that, in all but the prokaryotes and the simplest eukaryotes, the main means of evolving since the Proterozoic eon is by developmental change. Behavioural change may also have a role in the evolution of higher vertebrates, but there is little evidence for it as yet.

The second part of the book examines genetic changes with time: over a few tens of years (toxin resistance), several thousand years (selective breeding), several million years (differentiation of lower taxa) and many million years (formation of orders and classes).

Genetic resistance to toxins is usually a matter of metabolic adjustment, involving a limited number of minor changes because major changes are lethal. Nevertheless, many of the adjustments are large enough to be deleterious in the absence of the toxin, and represent compromises between death from toxicity and debilitation from the alteration of essential metabolism. Some groups like bacteria and insects readily become resistant to toxins because they can acquire (bacteria) or have partly evolved (insects) genetic systems for detoxification.

Evidence suggests that the practice of artificial breeding for economic purposes results in developmental changes mostly brought about by the selection of alterations in the hormonal control systems of animals and possibly of plants. It may be that the formation of species and genera also occurs in part by hormonal changes, but their origin has not been examined from that point of view.

The genetic processes concerned with the intermediate-term evolution of species and genera have been shown repeatedly to involve changes in gene regulation. The altered trans-acting factors implicated in plant and animal breeding are overshadowed in frequency and effect by the cis-located alterations of species differentiation (the terms 'cis' and 'trans' are explained in Section 2.2.8). Some cause morphological changes by altering the regulation of genes responsible for development, while others affect the developmental regulation of genes unconcerned with development itself. Apparently, alterations are usually in short flanking sequences of a gene which are sites for the binding of trans-acting factors in gene regulation. As a consequence, gene expression may be shifted with respect to

3

developmental stage, tissue or organ, or to environmental signals.

Marked differences exist in the evolution of prokaryotes, plants and animals at higher taxonomic levels. Since the prokaryotes have so little morphological development, they only serve to emphasise the paucity of evolution in its absence. The few aspects of spore development that some possess occur by a modification of the mechanism for metabolic adjustment to adverse environments; namely, an alteration in the transcriptional specificity of RNA polymerase. The plants have a simple form of development linked to environmental signals. They are characterised by a facility for morphological change that results in numerous species and genera, a low occurrence of pleiotropy and a high incidence of parallelism and convergence. There is no evidence that they have mechanisms for the evolution of orders and classes that are additional to those involved in plant breeding and in the formation of subordinate taxa. In the animals, in contrast, marked evolutionary changes appear frequently to be due to alterations in the morphogenic movement of cells, in cell adhesiveness and in induction-reaction processes during ontogeny.

The different means of generating genetic change are discussed in Part 3. There appear to be three broad classes: point mutations of one or a few nucleotides; alterations in the structure or number of chromosomes; and reshufflings of gene parts. The last class, which we can call loosely 'recombinational', occurs by processes of segment transfer, gene conversion, reverse transcription and transposition. As far as we can see at present, these three sorts of genetic variation may have had different roles in evolution: chromosomal changes seem to be of negligible importance because they rarely have advantageous phenotypic effects; point mutations alter genes slightly to make minor adaptive adjustments; and the formation of new or much altered genes probably depends on recombinational processes.

Part 4 of the book considers the evidence showing that the evolution of a group does not usually continue indefinitely. Each major group of organisms underwent an evolutionary radiation of novel characters sometime in the past, but many remained relatively static thereafter. The fossil record shows that stasis in innovative evolution usually means the termination of evolution in that group. Some static groups like the amphibians are declining, although environmental opportunities for novel evolution appear to continue. Others like the insects are very successful because they are in equilibrium with their environments and small ecotypic adjustments are sufficient to keep them there. However, nearly all groups are capable of producing species and genera by random events, which may be an inevitable consequence of a large population size.

The differences in evolutionary attainment in the various groups seems to be caused in part by a particular group adopting one adaptive strategy rather than another. When the potentialities of the chosen strategy have been exhausted, the group may have gone too far in one direction to shift to the strategy previously passed by. So the bacteria, by not acquiring or evolving mitochondria for efficient energetic metabolism, remain unicellular and microscopic; the fungi, by relying on absorption rather than photosynthesis, can exist only as parasites or saprophytes; and the plants, by retaining their cellulose walls, have shut themselves off from higher developmental and nervous systems.

The differences between groups in evolutionary potential may also be due in part to their having somewhat different mechanisms for the generation of genetic change. Major phenotypic characters, in particular, probably require the formation of one or more new genes by the assembly of parts of previously evolved genes. This is a process that seems to depend on the presence in the genes of nonprotein coding regions, or intervening sequences, of appropriate number, size and translational phase (see Section 4.5.6.5). Such intervening sequences are, however, absent in prokaryotes and most unicellular eukaryotes, they are much reduced in size in invertebrates and fungi, and in irregular translational phases in many of the genes unique to flowering plants. The unique genes of mammals, on the other hand, typically have more and larger intervening sequences of fairly uniform phase than do the genes common to vertebrates and other organisms.

The comparative evolutionary approach thus suggests that the focus of evolution has changed broadly from metabolism to development to nervous system. As the type of evolution has altered, so also has the mechanism, probably from mutational to recombinational. The development of a particular structure or function frequently blocks future innovative evolution, causing a decline in the evolution of many groups since the Cambrian.

5

PART 1
Succeeding Phases of Evolution

For nearly four thousand million years, evolution has been at work fashioning every aspect of an organism from its chemical composition to its nervous basis of behaviour. As we might expect, there have been successive phases during the history of life where evolution has concentrated on one aspect more than another. In part this has been due to the necessity of developing such attributes of life as metabolism as rapidly as possible in early stages, and in part it is because one feature of an organism must be evolved first to provide the basis for the production of another feature such as development. Consequently, the overall course of evolution may be divided into a number of different types or stages according to the property of the organisms acted upon, each with its individual mechanism and significance. Thus evolution may be biochemical (metabolic), macromolecular, subcellular, chromosomal, genomic, developmental or behavioural, some of which may be more important in certain episodes of evolution and in some groups of organisms than in others. Moreover, many of these evolutionary activities were essentially completed in the Proterozoic, allowing the nature and rate of the evolutionary process to alter markedly in the succeeding Phanerozoic eon (see Appendix 2 for the geological time scale). The major types of evolution are considered in the following pages, with the conclusion that the evolutionary focus since well before the Proterozoic eon ended has been mainly on matters of development, and perhaps latterly, in animals, of behaviour.

1.1 *Metabolic Evolution*

Although very important in the early establishment of life, metabolic evolution now appears to be of much less consequence. The basic genetic and enzymatic framework of energetic, synthetic and degradative reactions must have been established very early and has persisted with only minor change since its origin.

Much of this early evolution must have concerned the basic problem of energy supply. The evolution of an energy-generating system is essentially the development of a means of synthesising adenosine triphosphate or perhaps similar phosphorus compounds, so that synthetic and other activities requiring free energy can occur.

In their evolution during the Proterozoic, the prokaryotes developed all the main types of energy-yielding metabolism: fermentation, where organic compounds are both the donors and acceptors of electrons; anaerobic respiration, where the ultimate electron acceptor is an oxidised inorganic compound; aerobic respiration, where the ultimate electron acceptor is molecular oxygen; and photosynthesis, where light is the source of energy (Figure 1).

The initial course of biological evolution is thought to have been largely determined by metabolic inventions especially as regards the availability of oxygen. The first organisms to produce oxygen in their metabolism were the cyanobacteria which succeeded various forms of anaerobic heterotrophs, photoautotrophs and perhaps methanogens. This oxygen may first have been absorbed in the oxidation of inorganic compounds, but later it accumulated in the atmosphere. At an oxygen concentration near 1% of the present level, which may have been reached 1300 million years ago (Cloud, 1978), the evolution of efficient respiration became possible. The oxygen of respiration has probably been responsible for the extinction of many anaerobic organisms, for the dominance and variety of aerobic bacteria and eventually for the evolution of the eukaryotic type of cell. However, according to Morowitz (1971), the energy levels per unit mass of living material, whether bacterial cells or mammalian tissue cells, are constant at about five kilocalories per mole higher than the same material in the nonliving state. This implies than an increase in energetic complexity has not accompanied evolutionary advance.

The first known metazoa are found in the Ediacarian sediments with an age of nearly 700 million years. The thin flattened shapes of the Ediacarian worms are believed to have been necessitated by the short diffusion paths possible with low oxygen tensions (Raff and Raff, 1970). The lack of hard parts (shells, cuticles or carapaces) in these early faunas was perhaps because the formation of these structures requires collagen, and collagen synthesis needs molecular oxygen (Towe, 1970).

The other main aspect of metabolism is the synthesis of compounds necessary for the living cell. Although biosynthetic pathways are practically identical in all organisms in which they occur, they appear to have had several evolutionary origins. The little evidence available suggests that some of the more ancient syntheses have followed prebiological pathways by evolving the intermediates and the genes and enzymes responsible, backward from the original metabolite (e.g. the genes for aspartate and homoserine kinases in threonine biosynthesis; Truffa-Bachi et al., 1975). Some catabolic reactions which yield energy from intermediate stages may have evolved stepwise from the initial substrate to the final product (e.g. carbohydrate breakdown by fermentation; Degani

Energy Sequence		Electron Donors	Electron Acceptors
abiological energy sources	spontaneously formed organic phosphates	–	–
	↓		
anaerobic respiration	substrate level phosphorylation	free hydrogen or carbohydrates	oxidised elements or inorganic compounds later replaced by pyridines and still later by pyridine nucleotide (NAD)
	↓		
origin of electron transport	'oxidative' phosphorylation	reduced organic and inorganic compounds	electron transfer from NAD to flavoprotein and then on to porphyrins and eventually to oxidised inorganic compounds
	↓		
evolution of chlorophyll molecules from porphyrins of anaerobic respiration	photoassimilation	partly reduced inorganic compounds	quinone and cytochrome
	↓		
production of oxygen	photosynthesis	hydroxyl ions from the photolysis of water	additional electron carrier ferredoxin evolved to receive electrons from irradiated chlorophyll
	↓		
complete oxidation of organic substances to CO_2	aerobic respiration	organic and reduced compounds	electrons now transferable to oxygen

Figure 1 A possible sequence for the evolution of energy metabolism.

and Halmann, 1967). A third form of evolution, for pathways where the intermediate steps are without selective value, may have involved the duplication of genes with other functions, and their mutational alteration, to give enzymes of altered specificity catalysing a new metabolic sequence (e.g. the catabolism of certain aromatic compounds by oxidases which do not produce ATP for energy (Ornston and Stanier, 1966; Figure 2).

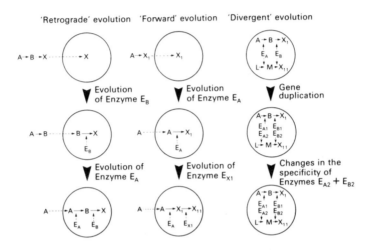

Figure 2 Patterns for the evolution of metabolic pathways. The circles represent protocells. Spontaneous reactions occur outside the protocells and dashed lines show the diffusion of extracellular substances into the protocells. X, X_1, X_{11} are metabolites required by the protocell; E_A, E_B etc. are enzymes evolved by the protocells.

Baldwin (1949) examined animal metabolism in relation to environment and concluded that 'there exists a common fundamental chemical ground-plan of composition and metabolism to which all animals, and very probably other living organisms also, conform'. In this connection, Britten and Davidson (1971) have pointed out that lower and higher organisms appear to have perhaps 90% of their enzymes in common. Also Medawar (1967) has remarked that he knows of 'no new kind of chemical compound that has come into being over a period of evolution that began long before animals became differentiated from plants. Nor has there been any increase in chemical complexity; no chemically definable substance in any higher organism, for example, is more complex than a bacterial endotoxin'. Apparently a common

9

biochemical heritage was established very early in the evolution of life. Since then, further biochemical evolution has mainly been concerned with variations in anabolism, as in the distinctive metabolic processes of blood flagellates, and in catabolism such as adaptive changes in the breakdown of nitrogenous waste compounds. Plants are exceptional in this respect for they synthesise a variety of secondary organic compounds usually from sugars, organic acids and amino acids; over 80% of all known organic compounds of natural origin have been isolated from plants (Swain, 1974). There is one aspect of biochemical evolution that may still be occurring frequently: modification of enzyme control systems, as shown by the great variety of regulatory controls in organisms that have an otherwise stereotyped metabolism. This is further discussed below (Section 2.3.1).

1.2 *Macromolecular Evolution*

Although evolution by the mutation and selection of nucleic acids and proteins for functional efficiency is very ancient, macromolecular sequences are still highly variable. Ayala (1984) has calculated, taking into account data from sequential electrophoresis, heat denaturation and peptide mapping, that the protein-coding genes of flowering plants are on average 27% heterozygous, those of invertebrates 22%, and those of vertebrates 28%; because of codon redundancy (see Section 3.1.2), the nucleotide variation will be even greater. There appears, however, to be no correlation between either reproductive isolation or rates of morphological evolution and degree or rate of change in protein sequence (Langridge, 1987). This conclusion accords with the proposition that the degree of amino acid difference is mainly a function of the time that the organisms containing the coding genes have been separated.

The proposition that amino acid substitutions can occur in a metronomic or regular manner has been called the 'molecular clock' hypothesis. It implies that a fraction of DNA nucleotide substitutions, including most third-base replacements in coding triplets and some of those altering protein sequence, is neutral or nearly so with respect to natural selection. The most direct test of this proposal would be to show that some at least of the alleles producing protein variants in a population are indifferent to selection, but for technical reasons this has proved difficult. A consequence of the proposition is that, following the origin of one group from another, their common genes and proteins should become progressively more different as time elapses. This prediction is supported by the finding that the degree of change in amino acid sequence in homologous proteins is approximately proportional to the duration, based on fossil evidence, that various mammalian species have been separated (Fitch and Langley, 1976; Figure 3). The existence of a molecular clock in the genome that acts to produce time-dependent divergence in protein molecules is reasonably well established. However, it is a clock that runs at different speeds for different proteins, and it can fluctuate in rate over long evolutionary periods.

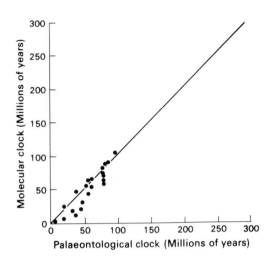

Figure 3 The mammalian molecular clock. The mean of the palaeontological range in millions of years for the origin or divergence of a group is plotted against the molecular clock date calculated from nucleotide replacements for seven polypeptides from globins, crystallins, fibrinopeptides, cytochromes and carbonic anhydrases. (Data of Goodman et al., 1982.) The molecular clock date of 174 million years seriously underestimates the fossil date of about 315 million years.

In addition to these continual genetic changes of the nature of point mutations, there are frequent chromosomal rearrangements and large-scale changes that occur in repeated DNA sequences, as discussed below. The potential evolutionary importance of these various genetic alterations and reorganisations is a matter of debate, but there is evidence that many of them are without phenotypic effect and are thus evolutionarily inert. The occurrence of these changes does not mean that significant evolution is still occurring at the biochemical or physiological level of sibling (morphologically identical) species, for example, nor it is necessary to invoke a 'cohesion of the genome' to maintain a constancy of morphology. Although 'genetic variation' as expressed in nucleic acid change is being continually generated, it is presumably not now of much evolutionary importance. It should not be expected that a high degree of macromolecular change is symptomatic of active evolution.

1.3 *Subcellular Evolution*

The interior of the prokaryotic cell has little apparent differentiation. Apart from ribosomes for protein synthesis and one or more 'nucleoids', the only discernible structures arise from ingrowths of the cytoplasmic membrane. These intrusions provide regions for DNA association, and thus orderly genetic segregation, for the location of light-absorbing pigment systems in photosynthesis and for the attachment of enzymes involved in respiratory function. Prokaryotic cells lack a distinct nucleus bounded by a membrane, have no mitotic apparatus, do not undergo meiosis; cytoplasmic streaming is absent and, although chlorophyll may be present, there are no organised plastids or mitochondria.

The genetic information of these cells was probably built up piecemeal by gene duplication or multiplication in prokaryotes or their predecessors during the Archaean and Proterozoic eons. These processes first generated the 500 to 1000 genes considered by Morowitz (1966) to be required to support a free-living organism, and then the 2000 to 3000 genes of present-day bacteria. An early evolution is expected of a semipermeable boundary

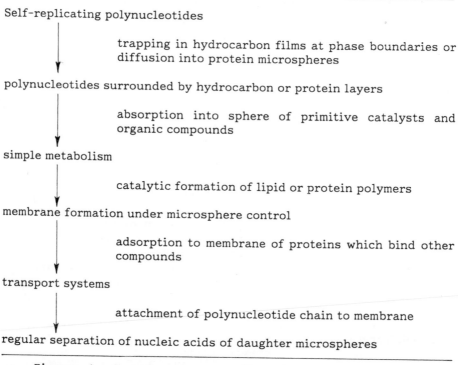

Self-replicating polynucleotides

⟶ trapping in hydrocarbon films at phase boundaries or diffusion into protein microspheres

polynucleotides surrounded by hydrocarbon or protein layers

⟶ absorption into sphere of primitive catalysts and organic compounds

simple metabolism

⟶ catalytic formation of lipid or protein polymers

membrane formation under microsphere control

⟶ adsorption to membrane of proteins which bind other compounds

transport systems

⟶ attachment of polynucleotide chain to membrane

regular separation of nucleic acids of daughter microspheres

Figure 4 A proposed succession of steps in the early establishment of the metabolising cell.

membrane in order to separate the replicating nucleic acids of the first 'organisms' from the mass of random molecules which form their precursors (Figure 4). But metabolism within the membrane-enclosed aqueous environment will tend to increase its solute concentration and the cell will be liable to osmotic bursting. This problem has required the evolution of mechanisms either to contain the cell at its higher osmotic pressure or to maintain it at an acceptable equilibrium with the external environment. The former means is provided by inextensible cell walls, as in plants, fungi and some bacteria; and the latter by the excretion of byproducts of metabolism of low molecular weight, as in animals.

The necessity to conserve energy and solvent capacity within the cell has required the evolution of a number of devices to activate gene expression only when intracellular conditions require it. In prokaryotes or viruses, several strategies for differential gene expression have evolved. They include the repression-induction system where an exogenous substance entering the cell releases the repression; sequential transcription of genes arranged in a linear sequence; alteration of nucleic acid secondary structure (hydrogen-bonded loops) to expose templates; and gene expression activated by earlier synthesised gene products. In order to maintain synthetic products at concentrations close to levels of utilisation, cells have evolved mechanisms to regulate the catalytic activity of many enzymes. To this end, selection has varied the stability of messenger RNA molecules and evolved multichain proteins which are sensitive to activators and inhibitors. Certain constancies of gene arrangement and patterns of DNA-folding suggest that even more fundamental means of regulating gene expression may remain to be uncovered.

An important aspect of intracellular substructure, first evolved in the prokaryotes, is the property of self-assembly; that is, amino acid sequences that have evolved to provide various self-determining protein shapes, either by folding or by specific associations (Figure 5). It is a means of converting one-dimensional information into three-dimensional structure, and this process appears to underlie much of the regulated activity and internal organisation of cells. It makes multimeric proteins like haemoglobin, multienzyme complexes like the pyruvate dehydrogenase system, respiratory assemblies containing 15 or more enzymes, bacterial flagella and, possibly in conjunction with other genetic information, organellar and cellular shape.

Above a certain size, a single genetic molecule, as in bacteria, is likely to be difficult to duplicate and segregate entire into daughter cells. Thus it became necessary to distribute the genes into separate chromosomes, as has occurred uniquely in all known eukaryotes. Correlated with the origin of chromosomes is the appearance of repeated nucleotide sequences, various patterns of sequence organisation and protein-DNA associations that were not present in the prokaryotes. For chromosome separation, mitosis evolved by a progressive development of microtubular structures to form a spindle apparatus with concomitant changes in chromosomal structure to provide sites (centromeres) for spindle fibre attachment.

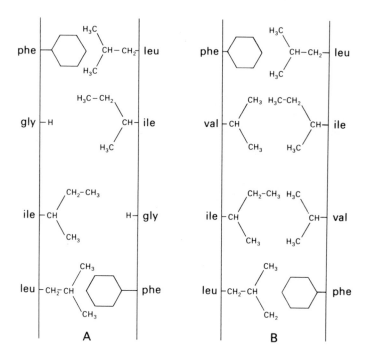

Figure 5 Steps in the evolution of protein association.
A. An opposing surface of two identical protein subunits aligned in antiparallel and held by two hydrophobic interactions between leucine and phenylalanine.
B. The same surface but with a further two hydrophobic interactions between isoleucine and valine brought about by a single mutational replacement of glycine by valine.

In sexual reproduction by cellular fusion there is a consequent doubling of the amount of genetic material in the zygotes. This two-fold (diploid) level must be reduced to the normal (haploid) amount if redundant DNA is not progressively to accumulate with each subsequent sexual fusion. The restoration of the haploid level has been accomplished by the evolution in eukaryotes of meiosis, which is apparently a derivative of mitosis (Stack and Brown, 1969). The process of meiosis is found in the Rhodophyceae (red algae) which are among the most primitive of eukaryotes, so it probably evolved very early in eukaryotic evolution.

The endoplasmic reticulum, only found in eukaryotic cells, is an extensive intracellular membrane network continuous with the cell membrane. It carries the ribosomes for protein synthesis and the enzymes for lipid formation, among others. It is sufficiently like the membrane invaginations of the prokaryotic cell for it to be probable that it is derived from this prokaryotic structure (Schnepf, 1964). Several proposals have been made for the origin of the nuclear membrane, but that of Pickett-Heaps (1974) suggesting its derivation from a part of an ancestral bacterial membrane which included an attachment site for DNA seems most plausible. This derivation suggests a continuity between the separation of newly synthesised DNA by growth of the plasma membrane in prokaryotes and a similar separation in early eukaryotes where the DNA was attached to the nuclear membrane.

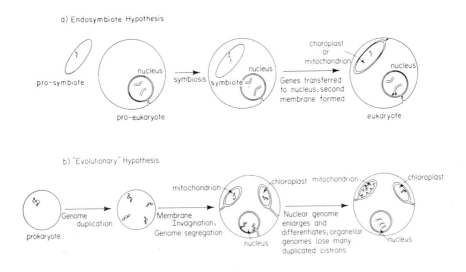

Figure 6   Two of the principal schemes for the origin of eukaryotic organelles (Buetow, 1976).
In the 'endosymbiote' hypothesis, a), a small prokaryote becomes established in a large prokaryotic cell where it becomes adapted, reduced and otherwise evolved to form an organelle.
In the 'evolutionary' hypothesis, b), some or all the duplicated genomes become enclosed in membrances and evolve differentially, to form, eventually, chloroplasts and mitochondria.

A postulated critical step in the formation of the eukaryotic cell was an alteration in membrane properties to allow the membrane to become detached in part and resealed. This feature may have enabled a new mode of nutrition by predation, and in particular the acquisition of what were to become the organelles (mitochondria and chloroplasts). Various proposals have been advanced for the origin of the multigenomic eukaryotic cell. One that has been presented in detail by Margulis (1970) is that invading organisms, if their genomes confer advantages on the host cell, may become symbiotic rather than parasitic, by the selection of mutant hosts which do not destroy the invaders. Such advantages are indeed conferred by the known extranuclear genomes: photosynthesis by chloroplasts and oxidative phosphorylation and other energy transformation by mitochondria. The overall similarity between chloroplasts and some cyanobacteria in both structure and function leaves little doubt of the endosymbiotic origin for chloroplasts (Figure 6). There are more difficulties associated with such an origin for mitochondria which vary so much in their amount of DNA in different organisms.

The evolution of intracellular organisation has been long completed, resulting in a close interdependence between structure and function. Nevertheless, cells of multicellular organisms are independently differentiated by programmed gene expression to perform different functions.

1.4 *Chromosomal Evolution*

Because changes that occur in chromosomes are sometimes thought to be an important component of the evolutionary process, they are considered separately from those of other subcellular structures. Changes in chromosome number by polyploidy and aneuploidy as well as alterations in structure by deletion, duplication, inversion and translocation constitute chromosomal change but not necessarily evolution. Certain of these 'mutations' may play a role in species formation, where they could act as fertility barriers, and in the occasional production of positional effects, but otherwise their contribution to higher eukaryote evolution appears to be slight. A correlation has been found between frequency of chromosomal change and degree of evolutionary divergence in animals (Wilson et al., 1975), but these chromosomal alterations are more likely to be a consequence of divergence than a cause of it. Polyploidy is common in plants where 40 to 50% of flowering species are estimated to be allopolyploids (Lewis, 1980), but it is rare in the higher animals, being certainly known only in some fishes and frogs (White, 1973). It tends to be associated in plants with apomixis, a type of reproduction without fertilisation, and thus may aid in the establishment of new species, but its other phenotypic effects are usually minimal. In some instances, there is evidence that polyploidy may be associated with changes in the ecological distribution of plant species, but if so the physiological or genetical reasons are unknown (Stebbins, 1971).

The actual number of chromosomes containing the nuclear genome also appears to be without significance. In two species of barking deer *(Muntiacus)*, the Chinese one has 46 chromosomes in the female and the Indian one only six. Yet the species are very similar morphologically, and somatic development is completely normal in the hybrid, indicating that it is irrelevant whether the developmental genes are in six or 46 chromosomes (Schmidtke et al., 1981).

16

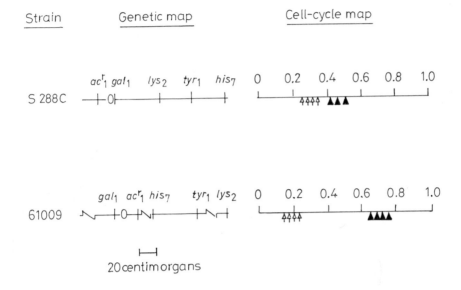

| Strain | Genetic map | Cell-cycle map |

Figure 7 The effect of gene position on the timing of enzyme synthesis. In two strains of yeast, the genes for α-amino-adipic acid reductase(△) and galactokinase(▲) are at different map positions on homologous chromosomes and the time of synthesis in the cell cycle is altered correspondingly (Halvorson et al., 1971)

There is evidence in prokaryotes that the order in which genes are located on chromosomes has evolved for purposes of functional efficiency. Some bacterial genera have retained the same gene order although the DNA has diverged widely in other respects. In *Salmonella typhimurium* and *Escherichia coli*, which are estimated to have diverged about 140 million years ago (Ochman and Wilson, 1987), 59 genes which occur in both genera have been located in the chromosome. Without exception, these genes occur at identical positions in the two DNA circles (Demerec, 1965). However, the nucleotide sequences of the DNA are very different because little recombination occurs in transduction hybrids (Demerec and New, 1965) and the DNA of the two genera reassociates poorly *in vitro* (Crosa et al., 1973). In yeast, sequential transcription of genes on a particular chromosome during a cell cycle is related to their linkage distance (Halvorson et al., 1971). The timing of gene expression can readily evolve and change in character merely by alteration of gene position. Thus, an increase in chromosomal distance between the genes for galactokinase and αaminoadipic reductase was shown to increase the difference in time of expression from 0.15 to 0.5 of the cell cycle (Cox and Gilbert, 1970; Figure 7). The conservation of gene order in bacteria and the effect of gene position on expression in yeast imply that rearrangements in chromosomal structure in these organisms many have large phenotypic effects.

Such effects of chromosomal rearrangement appear to be rare in higher eukaryotes, although there is evidence for synteny (location on homologous chromosomes) of many genes in mammals. For example, of 32 homologous genes mapped in the chromosomes of man and cow, 29 are found to be syntenic (Womack and Moll, 1986), indicating that even quite distally linked genes are normally kept together on the same chromosome. So, although the repeated sequences of DNA are in a continual state of flux as discussed below, the protein-coding genes are relatively stable in chromosomal position. The inference is that expression of these genes would be altered by translocation within the genome. On the other hand, experiments in *Drosophila* show that a gene may be placed in a large number of locations in euchromatic regions of the genome without altering the gene's normal regulation (tissue-specific and stage-specific expression), but in some instances its quantitative expression is altered (Wakimoto et al., 1986). In vertebrates there is no evidence for close linkage, at least for genes controlling metabolic proteins, even when their products act coordinately (Ohno, 1970).

There seems to be both inferential evidence for, and observational evidence against, there being an evolutionary significance to chromosomal change, but we can reasonably conclude that such changes are not usually important, except at the level of speciation in plants.

1.5 *Genomic Evolution*

The term 'genome' once referred to the haploid set of chromosomes in the cell of an organism, but now it more frequently means the total DNA content in the nucleoid of a prokaryote, the nucleus and mitochondria of a fungal or animal cell and the nucleus, mitochondria and chloroplasts of most plant cells. In this section, however, the word 'genome' refers only to the DNA of the nucleoid or nucleus.

As compared with flowering plants or vertebrates, the prokaryotes do not vary greatly in their amount of genetic information. Gillis et al. (1970) consider that bacteria with small genomes such as *Nitrosomonas, Moraxella, Hemophilus* and *Azotobacter* have limited enzyme activity. However, much of the eightfold variation in genetic information in bacteria may merely reflect differences in the ability to form enzymes adaptively; that is, only in the presence of an inducer. There is likely to be a set of genes, perhaps about 1000 as in *Mycoplasma meleagridis* with a genome size of only 0.42×10^9 daltons, which are present in all bacteria; they carry out similar and essential functions and are normally expressed under all conditions. In addition, a fraction of the DNA may only be expressed in particular environments, depending especially on the nature of the food available. This fraction will vary in constitution from one bacterium to another according to the nature of the environment it has evolved to inhabit, and in amount according to the range of environments and foods it has accumulated genetic information to deal with.

TABLE 1
Minimal haploid nuclear DNA contents in femtograms (10^{-15}g). With the exception of plants, there is a rough correlation between minimal genome size and complexity of the organism (data from Cavalier-Smith, 1978, and Sparrow and Nauman, 1976)

Prokaryotes	
Mycoplasmata	1.7
Other bacteria	3.3
Cyanobacteria	2.6
Eukaryotes	
Algae	40
Protozoa	60
Fungi	5
Plants	
Bryophytes	640
Pteridophytes	6000
Gymnosperms	4200
Angiosperms	1000
Animals	
Sponges	55
Coelenterates	330
Annelids	700
Arthropods	
Crustaceans	700
Insects	100
Molluscs	430
Echinoderms	540
Chordates	
Fish	390
Amphibians	
Urodeles	19000
Anurans	1200
Reptiles	1500
Birds	1700
Mammals	3000

The lowest reliably known eukaryotic DNA molecular weight is 3.2×10^{10} daltons in the haploid yeast, *Saccharomyces cerevisiae*, and the largest bacterial genome is 3.6×10^9 daltons in *Serratia marcescens*. In both of these genomes there is little repeated DNA. The larger genome in other eukaryotic cells is mostly due to repetitive DNA sequences, although part of it will be concerned with providing the extra genetic information necessary to specify the greater physiological and morphological complexity of the eukaryotic cell (Table 1).

Relatively recently it has been recognised that in plants, and particularly in animals, much of the DNA is in repeated sequences or 'families', as demonstrated by DNA reassociation kinetics. In man, for example, less than 5% of the nuclear DNA is in single sequences which code for proteins, while the rest of the genome consists of multiple copies of sequences of varying unit length and degree of repetition.

A small part of the fraction that is only moderately repeated comprises multigene families which code for ribosomal and transfer DNAs and such proteins as histones and immunoglobulins. Aside from these families, the function of most of this moderately repeated DNA is unknown. The organisation into multigene families appears to serve two functions. One is the production of gene products, such as ribosomal RNA and ribosomal proteins, rapidly and in large amount at certain developmental stages, as in the oocyte of the frog *Xenopus laevis* (Brown and Dawid, 1968). The other function of multigene families is to provide proteins such as the globin oxygen carriers for particular stages of ontogeny, or proteins that are required in different form in distinct cell types (Hood et al., 1975). In these cases, multiple related genes provide a means for the expression of slightly different genes to fit the peculiar conditions of successive ontogenic stages or cell types in the developing organism. Their nature and organisation will obviously differ among various plant and animal groups but too few have been examined for any evolutionary conclusions to be drawn.

A further fraction of the DNA genome, referred to as highly repeated or satellite DNA, consists of short sequences repeated a million or more times, which usually appear to be neither transcribed nor translated. While the single-copy sequences may be similar in amount in most mammals, the repeated sequences often vary greatly not only between higher taxonomic groups but even between species of a genus. The satellite sequences, in particular, undergo continual change in number, composition and position. These changes are thought to be brought about by the processes of unequal recombination, gene conversion and DNA transposition acting both within and between chromosomes. Such restructuring of nucleic acid sequences is regarded as 'genome evolution' although as yet no evolutionary role has been established for it.

The satellite sequences occur as several different families but within each family there is a marked conservation of sequence. Moreover, the satellite sequences of one species are often unlike those of another in both the sequences themselves and the number of repeats in a family. These facts have led Dover (1982) to suggest that there is a process of 'molecular drive' constituting a third force in evolution (the others being natural selection and genetic drift). His argument is that internal mechanisms of DNA turnover could bring about an almost synchronous genetic change throughout a family in a sexually reproducing population. If, as has been proposed (Rees, 1972), an abrupt but uniform alteration in DNA sequence can disturb chromosome pairing between individuals possessing the new sequences and the rest of the population, it would ensure reproductive isolation and incipient species formation. Alternatively, or in addition, if the new sequence or mutants of them had a phenotypic effect, there could be a gradual and concurrent change in all individuals of the population. Natural selection would be involved only to the degree that, if the phenotypic effect were disadvantageous, it would be selected against. The difficulty with this model is that instances are known where highly repeated sequences are almost identical in some species of a genus. *Drosophila gymnobasis* and *D. silvarentis*, for example, have undergone no significant change in either the amount or sequence of satellite DNA, although it comprises 40% of their genomes (Miklos and Gill, 1981). In this example, the evidence is against repeated sequence change having had a primary role in the speciation process. Moreover, highly repeated sequences have not yet been shown to have marked phenotypic expression. Indeed, one proposed explanation for the presence of highly repeated DNA is that, in the belief that it is functionless, it is 'selfish DNA' which, having the ability to multiply, transpose and insert itself into the genome, cannot be eliminated from the cell (Orgel and Crick, 1980; Doolittle and Sapienza, 1980).

The nuclear genomes of at least the higher eukaryotes are obviously in a state of flux as regards their repetitive sequences and perhaps have always been so. It cannot yet be decided whether this 'fluidity' of the genome confers adaptive advantages or whether it is forced on the genome as a byproduct of its structural organisation and the cell's enzymatic constitution. If the genome has a function in evolution other than that of providing the genic basis of phenotype, it has still to be demonstrated.

1.6 *Developmental Evolution*

An important type of evolution is that of development, as it is thought to be responsible for most of the progression that is observed now in multicellular eukaryotes. The process of development is usually subdivided into differentiation, or selective gene action, and morphogenesis, or the generation of shape. Differentiation leads to a progressive elaboration of cell structure and function, while morphogenesis may either be intracellular, specifying the three-dimensional structure of macromolecules, or involve the aggregation and redistribution of cells in tissue and organ formation. Both aspects of development are, however, usually closely integrated. In higher organisms, development at the molecular level comprises the appropriately timed synthesis of various morphogenic substances, their differential distribution, the establishment of gradients, and their interaction with other molecules, cell cytoplasm, cell surfaces and tissues. At the cellular level, the developmental

21

parameters include polarity, permeability, movement, adhesiveness and reciprocal interaction. Most of the developmental partial processes mentioned above have been shown to be genetically variable and thus subject to selective change.

Mutations affecting development in plants, which have much simpler processes of differentiation and morphogenesis than animals, frequently give viable functioning phenotypes. However, mutations altering components of early development in animals are almost always harmful, because later structures must often be built upon those formed earlier and also because early morphogenic stages may need to be retained because of their inductive roles later in ontogeny. It is a matter of common observation that new animal patterns of morphology arise usually by the modification of terminal stages of development. Consequently the degree or direction of evolutionary change is partly set by the development that has gone before, both in the ontogenetic and phylogenetic sense.

Although development of the familiar type is absent from prokaryotes and unicellular eukaryotes, these organisms possess a number of biochemical devices for gene switching. We shall see, in the discussion of developmental origins to follow, that some of them may have been elaborated to provide processes of development in multicellular organisms.

The origins of development

Just as there has been a number of independent origins of multicellularity (Figure 8), so has cell differentiation arisen anew many times. This process may have first evolved as a means of altering the uniform cell type to a form suited to withstanding severe environmental change. *Chlamydomonas*, for example, normally reproduces asexually when the cell divides into a number of small cells identical to the parent. But under conditions of nitrogen depletion, the cells form gametes which are differentiated from the parental cell in having sexuality and in the possession of fertilisation tubules; gametic fusion then produces a resting zygote with a thick wall (Friedmann et al., 1968). In this respect, some of the unicellular eukaryotes behave like many bacteria which form spores or similar structures when food becomes scarce or when the aqueous surroundings dry up. Whereas unicellular organisms express different sets of genes at different times, in their multicellular descendants different sets of genes are expressed in different cells.

When the constituent cells of an organism become interdependent and there is a consequent increase in body size, there is a change in the organisms's physical relationship to the environment, in certain respects similar to that of increased cell size. The volume of living matter increases as a cubic function of its linear dimensions, but its boundaries for environmental interactions such as gas exchange, heat loss, food assimilation and excretion only increase as the square function (Bonner, 1974). This discrepancy is thought to have led to selection pressure towards a differentiation within the cellular mass for purposes of metabolic maintenance. When estimates of numbers of cell types are plotted against

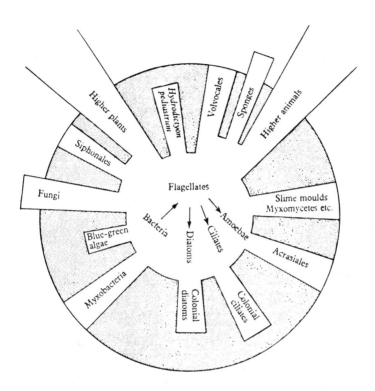

Figure 8 Diagram to illustrate suggested independent instances of the evolution of the multicellular state. Unicellular organisms are contained in the central ring and true multicellular organisms are shown projecting beyond the ring (Bonner, 1958).

the maximum sizes of organisms in various phyla and classes, there is a clear and fairly linear relationship between the two (Bonner, 1968). As with many other developmental processes, particularly of plants, the initial evolution of programmes of cell differentiation may be traced to the green algae. In existing algae belonging to the Volvocaceae, there is a progression from unicellular genera to multicellular colonial ones. Within this sequence there has been a gradual differentiation into two cell types, somatic and reproductive (Kochert, 1973). Subsequently the plants evolved a few more types of cell mainly in their adaptation to a land environment, while the animals, primarily because of their need for increasingly accurate movement in obtaining food, evolved several hundred types of cell.

Determination of shape in plants is mainly by alteration in cell division, with cellular differentiation being a consequence of *in situ* induction by hormones and other agents. These processes may have first evolved in the algae, because the contemporary green algae possess many of the structural and organisational features that eventually become fully established in the flowering plants. These include cell polarity, cellulose walls, multicellularity, shape by the control of the plane of cell division, cellular differentiation and apical meristems. Even the inducers of plant differentiation, the hormones, appear to have evolved early, perhaps in the prokaryotes (Greulach, 1973). Auxin has been found in algae and fungi, gibberellins in algae, and cytokinins in some bacteria and fungi.

In the multicellular animals, the arrangements that cells reach during development appear to depend upon mechanisms of cell recognition, migration, adhesion and reciprocal interaction. Some of these morphogenic processes also are already evident in the early eukaryotes. Thus cell migration, although probably not of the advanced animal type, is a feature of the slime moulds (Bueg et al., 1973) and hydroids (Hale, 1964); cell adhesion of some green algae (Bonner, 1974) and sponges (Müller et al., 1976); and cellular interaction or induction of the molluscs and ascidians (DeVilliers, 1965). It is likely that more complex animal development evolved by the modification of a few molecular strategies already present in the lower organisms. For example, Wolpert (1970) considers that the great increase in complexity from a sponge to a mammal has been made possible by a repeated use of only a few basic cellular mechanisms such as pseudopodial activity. For the bases of embryological processes, it has been remarked by Greenberg (1959) that zygotes frequently cleave to form hollow colonies in *Volvox,* or solid ones in *Synura,* which are morphologically equivalent to the hollow balls of cells forming blastulae. The same types of mechanism as are employed in the subsequent migration of the blastula cells (gastrulation) also occur in the algae and protozoa, for example, the invagination of daughter colonies of *Volvox* and the ingression of amoeboid cells in choanoflagellate colonies.

As to the number of genes or genetic systems in control of development, there is little information. Since proper development is essential to the survival and reproduction of the multicellular organism, it is expected that a large fraction of all inherited activity is concerned at numerous levels with the developmental process. In *Drosophila* it has been estimated by Raff and Kaufman (1983) that, of the approximately 5000 genes comprising the genome, the expression of about one-third is specific to successive developmental stages and is required for ontogeny to be correctly completed. Even simple developmental processes in the fungi need many genes for their determination. The number of genes specifically involved in the formation of conidia in *Aspergillus nidulans* has been put at 45 to 150 (Martinelli and Clutterbuck, 1971), and 48 ± 27 genes are estimated to be required for meiosis and spore formation in *Saccharomyces cerevisiae* (Tingle et al., 1973). However, Nüsslein-Volhard and Wieschaus (1980) could find only 22 loci in the *Drosophila* genome where mutant alleles produced alterations in segment numbers and polarity. For genes that act to provide positional information in *Drosophila*, it is estimated that there are unlikely to be many more than 20. The conclusion of Raff and Kaufman (1983) is that only 50 to 60 genes are required for the basic developmental programme for segmentation in *Drosophila*. These genes are considered to be concerned with the regulation of segmentation processes, not necessarily with the entire process of segment formation, which may require very many more. It does not appear to be known if intermediate alleles at these presumed controlling loci produce minor alterations in morphological form, as may be required for developmental evolution.

1.7 *Behavioural Evolution*

It would seem logical to suppose that the possession of organs receptive to ecological conditions, and an organ to analyse these perceptions and direct behaviour accordingly, would enhance an animal's use of the environment; that is, enhance its adaptability. Where the behaviour is exploratory, say of a new food source or a new ecological area, it may be of limited success with the animal's existing physical make-up. Then genetic changes affecting the morphological component of the activity may be selected for and, as the structure becomes improved for carrying out the activity, further behavioural changes may occur or evolve. If behaviour influences evolution in some such fashion as this, it should accelerate certain forms of evolution, especially aspects of adaptive morphology and physiology, over those occurring in the absence of behaviour.

TABLE 2

Brain size in relation to rate of anatomical evolution, as measured by change in body dimensions (Wyles et al., 1983). The correlation is more than 0.97, which the authors suggest may indicate that behaviour in higher vertebrates, rather than environmental change, is the major driving force for evolution.

Taxonomic Group	Relative Brain Size	Anatomical Evolutionary Rate
Homo	114	10
Hominoids	26	2.5
Songbirds	23	1.6
Other mammals	12	0.7
Other birds	4.3	0.7
Lizards	1.2	0.25
Frogs	0.9	0.23
Salamanders	0.8	0.26

This type of reasoning has been used by Wyles et al. (1983) to explain an observed correlation between brain size and evolution in vertebrates. They report a steady increase in rate of morphological evolution with increase in relative brain size from amphibia to man (Table 2). However, as we shall see, a consideration of nervous evolution provides little evidence either for or against an evolutionary role for behaviour.

1.7.1 *Origin of nervous systems and their influence on evolution*

The animal kingdom probably began with single-celled organisms which, while perhaps being photosynthetic, also ate bacteria to survive. In order to do this, the pre-zooflagellate would need to find its food, capture it and ingest it; each of these activities depends on specific responses to stimuli, requiring better receptive and effective mechanisms than ever evolved in bacteria or plants.

Some primodial nervous systems are present in contemporary organisms. For example, in the protozoan ciliates the cilia are coordinated by fine strands or neuronemes, but while the ciliate *Paramecium* has been reported to be able to learn (French, 1940), the neuronemes are merely submicroscopic conductive fibres and have not evolved further. Similarly, an attempt to evolve the rudiments of a nervous system occurred in the sponges which possess cells sensitive to specific stimuli but, perhaps because of a lack of cellular interdependence, no coordination or group control was possible.

The nerve net

It is only in one of the first known multicellular animals, the coelenterates of the Ediacarian period, that the basis of an evolving nervous system was laid down. Here the nerve cells are situated in masses within the body with extensions (dendrites) to the receptors to provide sensory endings and outgrowths (axons) to muscle or gland cells to provide a nerve net. But it is a poorly integrated net and, while it is effective in controlling movements of the body and tentacles, it is scarcely capable of determining behaviour. However, the invention of the neuron, including the cell body, sensory endings and axon, was a major step in the evolution of the nervous system because it enabled the whole animal to make a coordinated response to a stimulus and was retained with little modification throughout subsequent evolution.

Reflex arcs

A notable improvement on the primitive nerve net, where impulses can go in any direction, was the development of the synapse, a small gap where the branches of neurons meet, which ensures a one-way conduction of impulses and allows them to have an additive effect (positive and negative summation). With the evolution of the synapse, the basis of true memory is thought to have been established in the provision of reflex arcs (Elliott, 1970). The reflex arcs are channels of low resistance from sensory inlet to motor output, to enable a standard specific response, or reflex, to be made by the organism in response to a signal. The reflex arcs continue unchanged to the primates except that, as evolution progresses, they

26

become more under the control of the central nervous system, there is a tendency for the ranks of neurons to increase in number, and consequently the limits of adjustment and variation become broadened. Since reflex responses are not learned and are little affected by experience, they must be genetically determined and subject to mutational change. It is at this point that evolution acts to build different reflex arcs in different organisms to give appropriate stereotyped responses to factors of the environment. In the absence of reasoning, the reflex paths have an obviously high selective value.

The ganglionic nervous system

The next advance in nervous evolution was the development of the ganglionic nervous sytem which provides the stimulus-reaction response of all the invertebrate animals. The ganglia are separate collections of few or many neurons only loosely connected by nerve fibres. The advantage of having ganglia instead of a nerve net lies in the division of control that it provides, so there may be separate ganglia for the control of wings or jaws or for the senses like sight and smell. When under the control of a coordinating head or cephalic ganglion as in the flatworm, *Platyhelminthes*, the ganglionic nervous system has proved very successful because it is possessed by over 90% of animals. It has evolved in parallel fashion into numerous forms in different invertebrates, but has been unable to grow large, probably because the basic plan of cross-connections is too complex. Sometimes in invertebrates, the sensory organs are highly developed but the use of the information so gathered is limited by the ganglionic nervous system. In insects like bees, the eyes have developed well beyond the cephalic ganglia's capacity to analyse all the data received from them (Elliott, 1970). Consequently the ganglionic nervous system can provide for a variety of innate reflex responses but relatively little in the way of learned behaviour.

The various inventions comprising the nervous system of invertebrates (neuron, synapse, reflex arc, motor and cephalic ganglia) follow each other without marked discontinuity, but there is little knowledge of their genetic control or how they are affected by mutation. As judged by the mutations affecting neural development in the soil nematode, *Caenorhabditis*, most genetic changes in the nervous system have pronounced pleiotropic consequences, affecting both neural and nonneural development (Sulston and Horvitz, 1981). This suggests that radical changes to ganglion or brain structure would have effects as damaging as those interfering with morphogenesis. Behaviour, other than reflex, is minimal in the invertebrates, which means that it is unlikely that behavioural changes have much effect on the overall evolutionary process. However, it is evident that in general the more complex neuron organisation and the greater ranges of behaviour are found in the more actively moving animals (Romer, 1958). Since the ganglionic nervous system has been in existence for at least 700 million years, it is reasonable to suppose that little more is to be expected of it.

The brain

In the progression from invertebrate to vertebrate, or rather chordate, animal, some of the nervous components of the invertebrates, namely the neuron, synapse and reflex arc, were retained, but the organ responsible for a centralised nervous system, the neural tube, seems to have appeared suddenly. Among contemporary animals it is first found in *Amphioxus*, an animal believed to resemble the vertebrate ancestor. The neural tube represented an advance in structural organisation for it had the potential of growing to a large size without affecting, or being affected by, other growing organs. Although it did not give a functionally better coordination than some ganglionic animals possess, the neural tube provided a foundation from which all the other components of a central nervous system developed.

At the level of the fishes, the brain, composed of structures formed from the neural tube, consists of optic lobe, auditory reflex centre, cerebellum and hemispheres. Instinctive behaviour in the sense of inborn behaviour patterns, which are more plastic and intricate than reflexes, is believed to be conducted by the basal ganglia which first appear in fish at the base of the hemispheres in the brain. As judged by brain weight relative to body weight, there has been no evolution in the size of the brain in the Osteichthyes (bony fish) from the earliest Devonian to the present day, and only in the ray-finned Actinopterygii is there any sign of alterations in brain morphology reflecting changes in sensory perception (Jerison, 1973). This conservatism in brain evolution persisted through the amphibians and reptiles. Judging by versatility in behaviour, the average amphibian is slightly more intelligent than a fish, probably because of a greater level of coordination given by the development of the thalamus, and later by growth of another coordinating centre, the hemispheres. The impetus for these brain developments has been thought to have come from a change in sensory requirements on transference from sea to land. The usual conclusion is that, in the early lower vertebrates, the brain evolved to a certain level for the processing of information, and in later lower vertebrates there was some specialisation for types of information, without anything qualitatively new being added. In birds, on the other hand, there was not only a great increase in locomotor control with the adoption of flight, but the nesting habits of birds ensured a period of ontogeny when instinctive behaviour patterns could be developed before flight was possible.

With the advent of the mammals in the Mesozoic, the average brain size increased by four- or five-fold over that of amphibians and reptiles of comparable size (Jerison, 1973). Today, this level of relative brain size is still present in insectivorous mammals and North American opossums, but with the evolution of the more advanced mammals (ungulates, carnivores etc.) about 50 million years ago, there was a further four- to five-fold increase in brain size. In general, it seems that there was a rapid enlargement of the brain during the evolution from reptiles to mammals during the late Triassic, but little further change for about 120 million years until the later Cretaceous. Then the relative brain size of some orders increased steadily to the present, but there remained a great deal of diversity in size.

These trends and variations are usually attributed to selection pressure to adapt appropriate behaviour to particular environmental niches; and as the niches become more diverse, so does the brain size vary. Finally, the typical anthropoid brain became two or three times larger than that of the average mammal, culminating in that of man. Anatomically, the changes in the brain from fish to man have been very slight, with an increase in the number of neurons and further structural differentiation; from monkey to man, there was only a slight tendency towards continuing these evolutionary trends. Also, throughout nervous evolution there has been a steady progression in physiological characteristics such as greatly increased speed of conduction of stimuli and impulses, further integration of impulses, and increased dominance of sensory organs and motor control centres by the brain (Larsell, 1963).

Selective pressures
The evolution of the nervous sytem as outlined does not disclose anything new in terms of an understanding of evolutionary mechanisms because, even more so than in other types of morphological and physiological evolution, the nature of selected genetic changes is unknown. The selection pressures responsible for the continued evolution of the nervous system in metazoa are believed to have been, firstly, the necessity for controlled movement of tentacles and other food-trapping devices in filter-feeding animals, and secondly, the requirement to coordinate movement with sensory perception as food-trapping changed to active food pursuit. Major factors facilitating nervous system development and the evolution of behaviour may have been reproductive activities, the transition from water to land, the origin of homoiothermy (constant high body temperatures), flight in birds, and the evolution of the nursing habit leading to teaching and learning. As a consequence of the possession of a well organised nervous sytem, means were available for expressing inherited information relating to the environment to be encountered by the offspring; an organ for learning, memory and reasoning became available; and, finally, it enabled the faculty of speech in man. Since the intelligence of other mammals seems to be of the same nature as that of man, but in level only that of a three-year-old child, it is thought that the great difference in mentality between man and other mammals was brought about by the power of speech.

Behaviour and evolution
The real interest in the nervous system in the present context lies in the influence it may have had on the direction and rate of animal evolution generally. The effect, if any, of the nervous system on evolution is through its determination of behaviour which depends on sensory and motor abilities, but above all, on learning. It would seem likely that, as the nervous system has evolved, animals have concomitantly acquired greater potentialities for learning and for problem-solving, but this is very difficult to demonstrate. Comparisons of learning abilities in different taxa require a standard type of test and few are suitable over a range of

animals. However, learning experiments (conditioned reflex formation and maze learning) have indicated a more-or-less progressive increase in crude learning ability in invertebrates from worms to ants, but only the mammals among the vertebrates are more proficient in maze learning than insects (Warren, 1957). Among vertebrates, there appears to be little difference in conditioned reflex formation between fishes and primates, but reversal learning is very slow in fishes and rapid in primates (Voronin, 1962). Although in some reptiles there is a greater development of the cerebral structures, their learning abilities seem to be similar to those of amphibia and teleost fishes, indicating that despite expectations there was little if any increase in learning ability in going from sea to land (Harlow, 1958).

On the larger scale it is apparent that the complexity of behaviour generally parallels the degree of nervous structural complexity found in different animals. Furthermore, it has been observed that related species differ rather little in their behaviour patterns, which implies that much of the behaviour shown above the species level is an expression of the genome. Yet behaviour is much more subject to phenotypic change than are morphological and physiological features of an organism, presumably reflecting the learned component of a behavioural aspect. Little is known of the effects of selection on behaviour, but the consequences of behaviour have frequently been instrumental in causing reproductive isolation. In situations where related species inhabit the same area, it is desirable that they should not attempt to interbreed and that they should use different resources for food and breeding. Hence differences in behaviour with respect to the selection of mates, food, nesting sites etc. become important (Hinde, 1966), and much of the animal's behaviour is directed to these ends.

Since behavioural characters are brought about either by the expression of inherited information specifying or initiating a behavioural pattern or as a result of experience and learning, it can be asked why some are inherited and others learned. According to Sperry (1958), it is not because of a limitation in brain capacity, except perhaps for language, that some forms of behaviour are not inherited. A typical vertebrate brain may have about 10^{10} neurons, each with nearly 100 interconnections, but White et al. (1983) have concluded that there are insufficient genes to specify each connection individually. As it is, the nuclear RNA of the mouse brain corresponds to 42% of the single genes in the whole genome (Chikaraishi et al., 1978), requiring for its function much more genetic information than other organs like kidney or liver, where 10 to 26% of the genome is involved. This may mean that some information dealing with the conditions of life must of necessity be learned rather than inherited because, despite the brain's extensive capacity, the genetic basis for encoding behavioural information is inadequate. The complex interneural connections that probably constitute the integrative processes concerned with learning must therefore depend upon nongenetic adjustments or additions to the central nervous system.

Behaviour that is due to reflex action such as standing and walking soon after birth, suckling and pupil-response is entirely inherited in most mammals. Also, other forms of relatively invariable behaviour such as courtship, nest building and recognition, and pursuit of prey, are of such importance that they are primarily genetically specified. Where behaviour may need to be modified for adaptive purposes in the offspring, it would be expected that only the stimulus for a particular learning process would be hereditary. However, this probably depends on the attainment in the group of a sufficiently developed capacity for learning, and it appears that, in mammals except man, learning abilities are not well advanced. If circumstances allow the mammalian brain to evolve further, increased learning capacity will presusmably be favoured by selection over an amplification of instinctive behaviour patterns.

It is not possible to say precisely what effect behavioural variation or evolution has on evolution in general. In most instances morphology and behaviour have evolved or developed together in adapting an animal to its environment. That coordinate evolution is the desirable state is shown by occasional examples where one component lags behind the other. One such example in birds, where the behavioural aspect of an adaptive action seems to be inefficient or incompletely evolved, has been observed in seagulls (Tinbergen, 1960). The gulls break open shellfish with much labour by dropping them from the air, but they do not choose a hard surface for the drop, as crows do. Further instances of noncorrespondence between morphology and behaviour in social animals are given by Wilson (1975). In some other behavioural characters, it appears that the behavioural changes may precede the morphological ones. This is indicated when related species of birds utilise quite different structures in their courtship display movements and, by the tendency for the crown or neck feathers to evolve into crests in some species, to enhance a display that was presumably suboptimal (Lorenz, 1941). However, in general, behavioural and morphological changes seem to proceed in parallel as an adaptive complex and it cannot be said that behavioural evolution precedes morphological evolution, or the reverse. This being so, evolution in animals might be expected to be distinctly slower than in plants, because two complex systems, nervous and developmental, need to be adjusted in animals to make an evolutionary advance, but only a single simple one in plants. Perhaps in support of this expectation is the fact that, during much the same evolutionary period, the number of species of flowering plants has much exceeded that of animals (Section 2.3.2).

It is now believed that the ability to learn is provided by all nervous systems to varying degree. Particular types of elementary learning such as habituation, sensitisation and classical conditioning not only occur in both vertebrates and invertebrates, but take similar forms in each (Kandel and Schwartz, 1982). Consequently, it appears that learning, at least of the types mentioned, is of much the same nature at all animal phyletic levels. There seems to have been a gradual phylogenetic development of the learning capability, in which case perhaps morphological evolution should have increased in rate as animals have evolved. It is not known if this has been so.

1.8 *Conclusion*

The processes and direction of evolution vary in different groups depending on the feature of the organism that is subject to adaptive change, which in turn depends on its level of organisation. Certain aspects of evolution have virtually been completed long ago and, while the relevant genes still undergo mutational alteration, few permanent adaptive changes now occur in them.

The main phase of metabolic, including macromolecular evolution began before the first known fossil organisms 3500 million years ago; it probably became reorganised for aerobic metabolism when oxygen reached a suitable level about 1500 million years ago, and lessened well before the evolutionary emphasis changed to the development of intracellular structure (Table 3). Over the one to two thousand million years of biochemical experimentation, highly efficient processes of metabolic transformation must have evolved, because only minor alterations of them seem to have been made subsequently.

Since prokaryotes have little apparent internal structure, the impetus for subcellular differentiation was probably provided by the origin of the eukaryotic type of cell which is placed at or before 1400 million years ago (Schopf and Oehler, 1976). Intracellular evolution would be expected to be essentially over by the time multicellularity arose and differentiation into somatic cell types began.

The first multicellular eukaryote known is a possible macroscopic alga with an age of 1300 million years (Walter et al., 1976), but as the fossil is a carbonaceous ribbon, it may not have had differentiated cells except perhaps for spores or gametes. The oldest known metazoa are those of the Ediacarian period starting about 670 million years ago (Cloud and Glaessner, 1982), but Walter (1987) considers that the metazoa probably evolved one thousand million years ago. Judging from the complexity of the Ediacarian fauna, most members of which can be placed in extant phyla, this phase of cellular differentiation did not take long. However, it has remained an important evolutionary mechanism providing specialised cells for sensory, internal, muscular, and nervous organs, some of which are relatively late evolutionary innovations.

Morphogenesis is primarily an arrangement of cells into tissues to confer shape and functional structure which probably soon followed after cellular differentiation, that is, about one thousand million years ago. It now, or recently in the geological sense, seems to be more responsible for developmental evolution than is cellular differentiation. In combination with differentiation, morphogenesis has provided a central nervous system with the ability of directing animal behaviour. But it may be only in the higher vertebrates, and especially in man, that behaviour has had an influence on the direction and rate of evolution.

Like the origin of at least one of the organelles, much nuclear genome evolution in terms of mass seems to have occurred by saltatory events. There is evidence for total genome doubling in *Streptomyces and Escherichia*, and Sparrow and Nauman (1976) present data indicating a series of about seven doublings in progressing from bacteria to sponges.

TABLE 3
Minimal estimates of the time of occurrence of some events
before the Cambrian.

Eons	Years (10^9)	Geological and biological events	Evolutionary periods
Phanerozoic	0.0		behavioural evolution
	0.57	beginning of Cambrian	
	0.6	earliest metazoa in Ediacarian (Cloud & Glaessner, 1982)	developmental control
Proterozoic	0.67		(hormones)
	0.9 / 1.0	origin of animal phyla (Runnegar, 1982)	genomic evolution / developmental evolution
	1.3	O_2 concentration 1% (Cloud, 1978); earliest multicellular alga (Walter et al., 1976)	cellular interactions multicellularity
	1.4	earliest eukaryotes (Schopf & Oehler, 1976)	organelles
	1.6	oxygen at aerobic respiration level (Olson, 1978)	anaerobic to aerobic switch
	2.3	bacterial sulphate reduction (Walter, 1987)	metabolic evolution cellular structure
	2.5	oxygen-evolving photosynthesis (Olson & Pierson, 1986)	
Archaean	2.6		
	3.5	coccoid and filamentous anaerobic photoautotrophs (Dunlop et al., 1978)	
Hadean	3.7	oldest sedimentary rocks (Moorbath et al., 1973)	chemical evolution
	4.0	recrystallisation of earth's crust (Brooks & Shaw, 1973)	
	4.5	age of earth on lead isotope analysis (Ulrych, 1967)	

Thereafter the pattern is obscure, probably because of highly variable
amounts of repeated DNA in different organisms. This process of genome
doubling may have ceased by the early Palaeozoic because coelenterates,
molluscs, angiosperms and vertebrates all have much the same amount of
DNA in their basic genomes. Nevertheless, in their apparent number of
single copy genes, vertebrates with similar morphological construction may
differ by 10- to 100-fold (Britten and Davidson, 1976). A part of this
difference may reflect, not gene number, but variation in number and size

of intervening sequences which, according to Gilbert (1985) may constitute ten to thirty times the protein-coding DNA. Apart from intervening sequences which were already present when the plants and animals diverged about 1200 million years ago (Gilbert et al., 1986), there also appears to have been an increase during evolution of the protein-coding portion of the average gene. Kiehn and Holland (1970) found that the average polypeptide chain contains 190 amino acids in *Escherichia coli* and 235 amino acids in man, giving mean protein-coding sequences of 570 and 700 nucleotides respectively. There is additional evidence, based on rates of change of the same proteins in different species, that a great deal of total gene duplication occurred some 500 to 600 million years ago, before the appearance of the vertebrates (Doolittle, 1985).

With biochemical evolution being essentially over, macromolecular change being either neutral to selection or making slight variations to pre-existing sequences, and chromosomal, genomic and behavioural change being of uncertain significance, it may be concluded that the major evolutionary trends in higher organisms since the multicellular conditon arose have been developmental.

PART 2
Levels of Evolution

This part of the book sets out what we know of adaptive change at four distinguishable levels of evolution. The first two, the usually simply inherited metabolic changes that confer resistance to toxic agents and the artificial subspecific evolution that results from selective breeding, come under the heading of contemporary comparative evolution. The third is the formation of species and genera which seems to require changes in the cis-acting flanking regions of genes. The fourth is the formation of orders and classes, categories that require one or more developmental innovations in their origin. As might be expected, the nature of developmental change with the duration of evolution shifts from more probable and less significant alterations to rarer ones with greater phenotypic consequence.

2.1 *Contemporary Comparative Evolution*

The period from the advent of man as *Homo sapiens sapiens* about 34 000 years ago to the present is essentially a time during which many evolutionary changes have been due to human influence. Evolution, other than man-induced, has also been continuing in some organisms, but at an imperceptible rate; 34 000 years is only a tenth to a thirtieth of the time required to generate a species in rapidly evolving groups (Table 4). Only the activities of man in speeding up certain aspects of evolution, thus making trends detectable in a relatively short time-span, make contemporary evolution a topic worth considering. In essence then, a discussion of contemporary evolution is an account of the influence of man on his own and other organisms' evolution, which is contemporary in that it is still following the same trends.

Not all contemporary evolution, however, is relevant in a book on comparative evolution. The parts that are not are evolutionary changes confined to certain groups, such as the spread of viruses from wild hosts to man and his domesticates, and the genetic changes that may be occurring in human populations. Nor does microbial breeding, involving as it does the

TABLE 4

Estimates of the times taken for species and genera to produce, or to be transformed into, new species and genera.

Organisms	Species duration (years x 10^6)	References
Scaphitid ammonites	0.5–1	Simpson, 1953
Pelycodus (extinct primate)	1	Brookstein et al., 1978
Hyopsodus (archaic ungulate)	0.3 – 0.7	Brookstein et al., 1978
Agropecten (extinct mollusc)	5	Waller, 1969
Clams	1.25 –7	Stanley, 1976
Average invertebrate (marine)	6	Durham, 1971
Average mammal (Pleistocene)	1.2	Stanley, 1976
Average mammal (Quaternary)	0.33	Kurtén, 1959
	Genus duration	
Clams	70	Van Valen, 1973
Rudists (Mesozoic clams)	20	Van Valen, 1973
Mammals	6	Van Valen, 1973
Average mammal (N. American Pliocene)	5	Webb, 1969
Average mammal (N. American Pleistocene)	3	Webb, 1969
Horses	7.5	Simpson, 1953
Rodents	5.9	Gingerich, 1976

selection of appropriate metabolic mutants, differ much in nature from the mutational steps that allow the utilisation of new carbon and energy sources (Section 3.1). There remain two aspects of contemporary evolution which are informative from the comparative point of view. One is the evolution of resistance to synthetic toxic agents, which is important in the adaptation of bacteria, fungi, plants and insects to the suppressive activities of man. Although the use of toxins against unwanted organisms is a fairly new human activity, the poisoning of competitors or predators is an ancient evolutionary tactic of certain groups which may therefore have already laid the foundations for pesticide resistance. The other feature of contemporary evolution of comparative interest is the artificial selection of plants and animals for economic ends. This subject has been generally neglected as a source of evolutionary information despite the view of Haldane (1954) that 'one of the most hopeful fields for the study of evolution is the domestication of animals and perhaps also of plants'.

2.1.1 *Evolution of resistance to toxic agents*

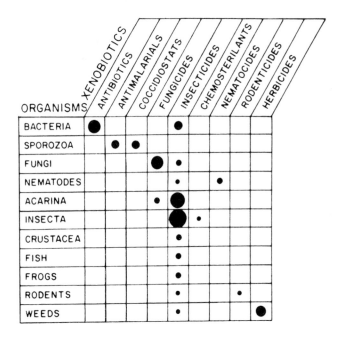

Figure 9 The relative frequency of resistance to xenobiotics (Georghiou, 1986).

An inadvertent evolutionary activity of man that is becoming increasingly important results from his attempts to kill or inhibit organisms that cause disease in, or compete with, man or his crops and herds. Organisms considered undesirable range from viruses to mammals, but most effort is exerted against bacteria and fungi pathogenic to plants or animals, weeds that compete with crop plants and insects that eat plants or spread disease. Without pesticides, for example, world production of food would be reduced by an estimated 30% (Barstad, 1978).

It is usually desirable to suppress the pest without harming adjacent organisms by directing inhibitory or lethal chemicals against features of structure or function that are peculiar to the target organism. However, severely inhibiting compounds that act selectively need to be analogous to some cellular component of the pest. An almost invariable consequence, because of the very high selective pressures generated, is the development of genetic resistance (Figure 9). Certain basic mechanisms of resistance, generally at a low level, are possible to all organisms, but different groups vary considerably in their capacity to evolve chemical resistance at high levels of inhibitor. The study of resistance developed to inhibitory chemicals can therefore provide information on metabolic evolution, revealing something of its mechanisms, its genetic origin and its variation among organisms.

Bacteria

Because many bacteria are pathogenic for man, they have been subjected to intense antimicrobial treatment, first by antiseptics, and since about 1940 by antibiotics. The most commonly used inhibitors of bacteria are the antibiotics, that is, antibacterial agents made by other organisms, usually species of *Bacillus*, actinomycetes or fungi. Since they are biologically produced to act against bacteria, bacterial selectivity is usually built into the structure. Thus some compounds inhibit cell-wall biosynthesis (penicillin), interfere with cell membrane functioning (chlortetracycline), affect the metabolism of DNA (actinomycin) or RNA (rifamycin) or block protein synthesis (chloramphenicol). Biochemical differences between bacteria and most other organisms in the enzymology of nucleic acid and protein synthesis or in the nature of walls and membranes ensure that these antibiotics primarily affect bacteria.

When bacteria are grown in isolation, as in the laboratory, genetic resistance to an antibiotic may arise by point mutation, perhaps with a frequency of 10^{-6} to 10^{-9} per generation (Braun, 1965). Such mutations often either decrease the entry rate of the antibiotic by altering cellular permeability (Chopra and Howe, 1978) or they alter the target site in the affected enzyme (Gellert et al., 1976) or ribosome (Cundliffe, 1980) so that the antibiotic no longer binds. The disadvantages of these resistance mechanisms are that a reduction in permeation usually gives resistance only to low concentrations of the toxin, and mutants with altered binding sites are often at a growth disadvantage in the inhibitor's absence (Kiser et al., 1969).

Eubacteria have coexisted on earth with the actinomycetes for at least

600 million years, and with the fungi for perhaps 1000 million years. Even if antibiotics against bacteria have not been produced for so long, there remains an extended period for antibiotic resistance mechanisms to develop in bacteria. As a result, genes have evolved specifying highly efficient detoxification mechanisms together with genetic means for their wide transfer, which are known to have been present at least 100 years before man began to use antibiotics against bacteria (Kiser et al., 1969). The evolution of these resistance systems was not due to the actions of man; his influence has been in provoking their extensive distribution by his medical and agricultural use of antibiotics. The genes for resistance usually form part of a mobile element (transposon) carried on a self-replicating DNA circle (plasmid) capable of cell-to-cell transfer by conjugation, transduction and transformation (Section 3.3.1). In order for resistance to be immediately conferred on the recipient cell, the resistance gene should have a dominant expression which is most easily achieved by the specification of a detoxifying enzyme.

Under strong selective conditions, such as provided, for example, by a high concentration of an antibiotic, the cell may survive by increasing the number of copies of an existing gene coding for an antibiotic-inactivating enzyme. An increase in template number generally leads to a proportional increase in the amount of enzyme synthesised, and there is a linear correlation between the concentration of modifying enzyme and the amount of antibiotic inactivated. When antibiotic resistance genes are carried on plasmids, a type of gene amplification leading to an increased number of gene copies often occurs by an increase in plasmid number per cell (Nordström et al., 1972) or in number of gene copies per plasmid (Yagi and Clewell, 1976).

Fungi

Most plants growing in the wild are genetically heterozygous, phenotypically variable and seldom free of disease. But when grown by man as crop plants, they are required to be high-yielding, homogeneous and resistant to disease. For the last reason, anti-plant-parasitic agents are most commonly directed against pathogenic fungi.

In order that the fungal parasite is inhibited but not the host plant, fungicides are designed so as to affect metabolic processes unique to the fungus. Examples are: polyoxin antibiotics which interfere with chitin synthesis in the fungal cell wall, the plant wall being of cellulose and therefore not affected; benzimidaxole compounds which affect the assembly of fungal but not plant tubulins; and pyrimidine derivatives which inhibit sterol synthesis in organisms containing ergosterol (fungi) but not those with lanosterol (plants).

Mutations conferring resistance to most fungicides are relatively rare, especially to multi-side inhibitors like dithiocarbamates and metal toxicants. Where resistance mutations do occur, they are usually ones which alter the target site to decrease the binding of the inhibitor (Davidse, 1982) or ones that reduce its transport through the membrane (Hori et al., 1977). Detoxification has rarely been reported as a mechanism of fungicide resistance (Dekker, 1986), although it is a common resistance mechanism in insects. Natural antibiotics or phytoalexins are produced by plants in response to infection and, although less inhibitory to fungi than synthetic fungicides, are frequently modified by fungal metabolism. Most of the known conversions involve the creation of new hydroxyl groups by hydration, carboxyl reduction, ether cleavage and particularly by oxygenation (Van Etten et al., 1982). It is doubtful, however, if these conversions represent the action of a detoxifying mechanism because the derivative is often unchanged in its fungal-inhibiting property. Where the modified phytoalexin is less inhibiting, it may be a chance consequence of the metabolism by the fungus of a plant-produced compound.

The inability of the fungi to develop detoxifying enzymes by mutation is only partly explained by the multi-site action of some fungicides. The fungi seem to be somewhat deficient in detoxification mechanisms for phytoalexins and accordingly do not have a metabolic basis for generating fungicide resistance by detoxification.

Plants

As an economical alternative to manual or mechanical weeding of crops, man has synthesised a variety of compounds (herbicides or weedicides) to inhibit selectively the growth of non-crop plants. As a first level of discrimination, the herbicides are directed against photosynthetic processes of the type that only occur in plants or against the synthesis of those amino acids that plants make but animals cannot. The second level of discrimination is between crop and weed plants, and this occurs fortuitously as a result of differential uptake or detoxification of the herbicide. For example, atrazine inhibits photosynthesis in many plants by binding to the chloroplast Q_B protein, but maize and sorghum inactivate it by an endogenous glutathione transferase. Accordingly, atrazine has been widely used to control weeds in fields of maize and sorghum. Atrazine, like most herbicides, has single-site action and, as a result, resistance to it has developed in several species by alteration of the target site. Most of the resistant mutants, however, have somewhat impaired photosynthesis so that their fitness in relation to the susceptible variety is low in the absence of the herbicide (Holt et al., 1981).

Resistance to the inhibitors of amino acid biosynthesis is believed to take much the same form in plants as it does in bacteria which have the same synthetic pathways. Mutant bacteria that are resistant to such herbicides usually have an altered target site in which a single amino acid has been substituted; e.g. in sulphometuron resistance in *Escherichia* (Yadav et al., 1986), glyphosate-resistance in *Salmonella* (Stalker et al., 1985), and probably glyphosate-resistance in tobacco cells (Singer and McDaniel, 1985). Resistance to inhibitors of amino acid biosynthesis, but seldom to photosynthetic inhibitors, can also arise by amplification of the gene for the target enzyme. Thus glyphosate-resistance may be attained by overproduction of the sensitive shikimate synthase in plant cells (Amrhein et al., 1983) and in *Escherichia* (Rogers et al., 1983).

In plants, as with fungi, the mutational production of a new or altered enzyme which detoxifies the inhibitor is not known to have occurred. This is despite the fact that the natural resistance of some plants, which permits the selective use of certain herbicides, is based on an endogenous detoxifying enzyme. It is not known whether time has been insufficient to evolve such detoxifying means of resistance by point mutation, requiring as it does a new or much altered enzyme, or whether such evolution requires mechanisms other than point mutation.

Insects

The greatest effort exerted by man in pest control has been against insects, but the insects have coped a great deal better with the onslaught than have most other pests. This is shown in the figures for average crop losses in the USA (May and Dobson, 1986); in about 1946 7% was lost to insects, but by 1984 this loss had increased to 13%. This is despite the fact that, of the four basic chemical classes of insecticide (chlorinated hydrocarbons, organophosphates, carbamates and photostable pyrethrins), all but the chloro-compounds were introduced after 1946. Unlike plant pathogenic fungi and herbaceous weeds, the insects have acquired resistance to toxins extraordinarily rapidly. The most significant factor in ensuring insecticide selectivity is the singularly easy penetration of the insect integument by insecticides. Such toxins as benzene hexachloride, pyrethrins and DDT are almost as toxic to mammals as to insects when injected, but whereas mammals are little affected by external application, insects are inhibited similarly whether the insecticide is applied superficially or injected (Winteringham, 1957).

As with other organisms, resistance occurs by target site alteration and by decreases in uptake, but the most usual mode of insecticide resistance is by an increased capability for detoxification. Many insects possess catabolic enzymes such as esterases or enzyme systems such as microsomal oxidases of low substrate specificity which can be mutated to detoxify synthetic insecticides. Sometimes the mere amplification of an unchanged hydrolytic esterase brought about by gene duplication confers adequate resistance. Such gene duplications have probably occurred during the development of aphid resistance to organophosphates (Devonshire and Sawicki, 1979). But in many instances, the resistant insect possesses an enzyme with an increased specificity for the toxin (Ottea and Plapp, 1984). It is not clear, however, whether the mutation conferring resistance has altered the existing enzyme or has rendered inducible or constitutive a latent gene for the detoxifying enzyme.

The numerous enzymes having detoxifying activities were evolved by many insects long before synthetic insecticides came into use, as is evident in their well-developed genetic and biochemical determination. Many of them are inducible by foreign toxins and the induction is often coordinate, inducing a range of detoxifying enzymes (Terriere and Yu, 1973). Moreover, there is evidence that suggests that some insects may possess receptor proteins which recognise and bind insecticides, and presumably dietary toxins, and transport them to the nucleus where they induce the detoxifying enzymes (Plapp, 1984).

41

TABLE 5

Detoxification activities of midgut microsomal epoxidase in last-instar larvae of species belonging to several families of plant-eating Lepidoptera according to the range of host-plant families normally eaten. Epoxidase activities are expressed as nanomoles of dieldrin produced by the oxidation of aldrin per mg of gut protein per minute times 10^3. Since significant differences in epoxidase activity are not found on rearing larvae on different food plants, the tabulated figures probably represent genetic differences between species and not differences in enzyme induction (Data from Krieger et al., 1971).

Lepidopteran Family	No. of Lepidopteran species	No. of families of plants eaten		
		1	2-10	11 or more
Saturniidae	6	–	70	60
Lasiocampidae	2	80	–	108
Geometridae	1	–	34	–
Sphingidae	1	3	–	–
Notodontidae	1	32	–	–
Lymantriidae	1	–	–	175
Noctuidae	17	10	110	475
Arctiidae	4	–	33	314
Nymphalidae	1	11	–	–
Danaidae	1	8	–	–
Mean epoxidase activity		20	91	294

It seems that many insects have such a highly evolved detoxification system because of the toxins they encounter when feeding on plant tissues or putrefying animal tissue. Plants elaborate a wide range of inhibitory chemicals such as phenols, alkaloids, terpenes and tannins, while decaying meat contains such potent toxins as cadaverine and putrescine. Lepidopteran pests of tobacco, which contains the alkaloid nicotine, are resistant to nicotine poisoning (Dyte, 1967). Furthermore, detoxifying enzymes like mixed-function oxidases and glutathione s-transferase are induced in Lepidoptera when feeding on plants (Yu, 1982) and the microsomal epoxidase activity of plant-feeding Lepidoptera is correlated with the range of plants eaten (Table 5). Accordingly, it would be expected that insects that feed on crop plants would frequently develop insecticide resistance by detoxification, as is indeed the case. In contrast, insects such as ticks, which feed on the blood of animals, would be expected to have few detoxifying enzymes and to attain insecticide resistance by other means. Nolan (1986) has reported that resistance to insecticides is usually due to an alteration in the binding site of the target enzymes in ticks and mites, but to detoxification in Diptera.

Origins of resistance
The way in which resistance to toxins has developed in various organisms illustrates two aspects of the evolutionary process at the metabolic level. The first concerns the relative frequency and nature of genetic change to adapt to metabolic stress and the second is the influence of past evolution on the nature of the genetic adaptation.

When a large population of susceptible organisms is uniformly exposed to an inhibitory or toxic chemical, a few individuals will be partially or wholly unaffected because of the spontaneous random mutations they carry. Such mutations may confer resistance by decreasing permeability, duplicating genes, altering sensitive nucleic acids or proteins, inactivating enzyme systems, activating bypass pathways or blocking ones increasing toxicity, reactivating cryptic genes etc. If the toxin is present at sublethal levels, one or other of these mutant single genes is likely to give sufficient resistance to allow reproduction, and with sexual or other means of recombination, a multigenetic basis for full resistance will develop. This is what is found in bacterial resistance to chloramphenicol (Cavalli-Sforza and Maccacaro, 1952), fungal resistance to ergosterol inhibitors (Van Tuyl, 1977), triazine resistance in *Senecio vulgaris* (Holliday and Putwain, 1980) and blowfly resistance to diazinon (McKenzie et al., 1980). Presumably because the accumulation of several resistance genes affects fitness, this type of resistance, which is often only to low toxin levels, is liable to decay when selection pressure is removed.

The survivors of a lethal or strongly inhibiting level of toxin are usually single-gene mutants which most commonly have an altered target site or have the relevant gene amplified. The mutations most frequently conferring resistance are inactivations of genes concerned with toxin entry and transport, those converting the compound to a more toxic one and those inactivating induction systems. Although mutations reducing the permeability or uptake of a toxin seldom confer sufficient resistance by themselves, they help to reduce inhibition and are frequently present in organisms with other resistance mutations. Mutations duplicating or amplifying a gene are less frequent, and those altering the target enzyme so that it still functions but no longer binds the inhibitor are even less common. Evidently, rather precise amino acid substitutions are required to be made in these instances. The production *de novo* of a detoxifying enzyme by mutation is rarely if ever encountered, although an existing gene may be mutated to change the enzyme's specificity slightly so that it can detoxify the inhibitor. An alternative mechanism of resistance, available only to the insects and other animals with nervous systems, is behavioural change. It appears to depend on mutations in signal receptors or signal processing conferring a hypersensitivity to low concentrations of insecticide so that avoidance can be taken before lethal effects ensue

43

(Georghiou, 1972). In general, an examination of acquired resistance mechanisms shows the same limited scope for spontaneous mutations of the point mutation type as that discussed later in relation to the utilisation of new substrates in bacteria.

It is notable that the frequency with which resistance is acquired, and some of the types of resistance attained, differ markedly among the groups examined. Bacteria and insects rapidly acquire resistance by detoxification, but fungi and plants are much slower to become resistant and, when they do, they rarely detoxify the inhibitor (Table 6). Gressel and Segel (1978) have remarked that antibiotics, chlorinated insecticides, dicoumarol rodenticides and herbicides were all brought into use about the same time, yet resistance soon appeared to all but herbicides. The generation time of weeds is usually longer than that of other pests; nevertheless, as many generations have been completed in plants since herbicides were used as were taken to develop insecticide resistance.

TABLE 6

The types of mechanisms used by various organisms to achieve resistance to toxins. + = of common occurrence; $\pm$ of rare occurrence; ? of doubtful occurrence.

Resistance mechanism

Organism	Altered target site	Decreased permeability	Detoxification	Gene amplification
Bacteria	+	+	+	+
Fungi	+	+	$\pm$	$\pm$
Plants	+	?	$\pm$	+
Insects	+	$\pm$	+	+

In most instances, stable bacterial antibiotic resistance is not acquired by mutation but by plasmid-transfer processes. Nevertheless, mutations may be needed to increase the substrate affinity of plasmid-determined enzymes, especially for chemically modified antibiotics. Considering the apparent difficulty of making entirely new enzymes by point mutation, the antibiotic-modifying enzymes probably originally evolved in antibiotic-producing bacteria for the prevention of autotoxicity. In support of this view of their origin, Benveniste and Davies (1973) have demonstrated that certain modifying enzymes from actinomycetes that produce antibiotics

44

catalyse precisely the same reactions as do the enzymes of antibiotic resistance factors. Moreover, the gene in the bacterial transposons 5 and 903 for aminoglycoside phosphotransferase, which inactivates the antibiotic neomycin, shows significant nucleotide homology to a gene for aminoglycoside phosphotransferase in *Streptomyces fradiae*, a producer of neomycin (Thompson and Gray, 1983).

The insects seem to owe their ready development of resistance by enzymatic detoxification to the array of detoxifying enzymes they already possess as a result of their feeding habits. Thus, although the acquisition of resistance requires an alteration of the genome, many insects are to a large extent preadapted to its attainment when toxins are encountered. This phenomenon is an example of how preceding evolution is instrumental in determining later pathways of adaptation.

2.1.2 Domestication and breeding of plants and animals

In writing *The Origin of Species*, Darwin (1859) arrived at evolution by natural selection by asking the question 'can the principle of selection, which we have seen is so potent in the hands of man, apply under nature?' The question is reconsidered here in the light of modern knowledge to assess whether artificial selection gives any indication of the processes of natural selection. The genetical and physiological changes that occur during artificial evolution have been subjected to a much more detailed study than have those occurring during natural evolution.

The survival of mankind in all but a small fraction of his present numbers is utterly dependent on cultivated plants and domestic animals. By processes of domestication, chosen wild organisms have been made amenable to the practices of man, and by selective breeding those characters of interest to him have been altered or enhanced. Vavilov (1951) called these processes 'evolution directed by the will of man'. Artificial selection is exerted with an intensity up to about 10 000 times that of natural selection, but it presumably utilises similar sorts of genetic variation. The major difference is that, whereas natural selection operates via differential reproduction, artificial selection is often exerted prior to reproduction. The relation between man and his domesticates is essentially a symbiotic one, for man could not maintain his numbers without crops and herds, and most fully domesticated organisms could not exist without man's protection. In some organisms like the banana and the breadfruit, and the camel and the dog, the only forms now known are the domesticated or feral ones.

The first organisms known to be domesticated for food were the cereals, barley and wheat, and the artiodactyls, goats and sheep (Frankel and Soulé, 1981). As discussed below, the processes of domestication are rather different in plants from those in animals; plants are usually roughly fashioned to cultivation by the selection of single gene mutants, but many animals require multigenic changes in behaviour before they become docile. Methods of selective breeding also differ between plants and animals because of differences in number and value of individuals screened for selection, variations in breeding systems (self- or cross-fertilised) and the possibility of somatic reproduction in plants. In both groups of organisms, breeding for economically important characters results in an increase in homozygosity unless steps are taken to preserve hybrid vigour, unlike the highly heterozygous constitution of most wild populations.

Plant domestication and breeding
   Most of man's food is derived from a small number of cultivated plants. According to Heiser (1973) almost 200 species of plant have been domesticated for food, but only about 15 of these are presently important. Mangelsdorf (1975) lists them as wheat, rice, maize, sorghum, barley, beans, peanut, soybean, sugarcane, sugarbeet, sweet potato, manioc, potato, banana and coconut.
   The main source of food lies in the reproductive organs of plants, many of which produce significant quantities of grain, seed, fruit and tubers in the wild state (e.g. Harlan, 1967). They were gathered by Neolithic and earlier man, but it was only when some of the gathered seed was resown about 10 000 years ago that man began to change the genotypes of the plants. The unconscious selection exerted by the processes of harvesting and replanting provided the first steps in domestication. Some of the characters adapting the plant to wild conditions are incompatible with those required in a primitive form of cultivation, but they are usually very variable because of the generally heterozygous and segregating nature of wild populations. As Pickersgill and Heiser (1976) have reported, a field of a high-producing cereal shows much less variation in characters like height, number of nodes, time of ear initiation etc. than does a comparable population of the wild form.
   Ecotypic adaptations to natural conditions, which are unsuitable in the cultivated state, include features facilitating seed dispersal, like fragile rachis, retentive glumes and awns in cereals or dehiscent pods in legumes and crucifers, and protective mechanisms like bitter or toxic principles, thorns and prickles. Although most of these characters are specified by many genes, single recessive mutations will usually reduce or abolish the adaptation. Examples are brittle to tough rachis in barley (Harlan et al., 1973), delayed pod dehiscence in legumes (Gladstones, 1967), spreading- to side-panicle in oats (Garber, 1922), awned to awnless in wheat (Watkins and Ellerton, 1940), high to low coumarin in *Melilotus* (Stevenson and White, 1940), spiny to smooth surface in cucumbers (Strong, 1921) and tall to dwarf stature in numerous crop plants. These single-gene mutants would be expected to be among the first selected by primitive man; this is supported by evidence from archaeology (Kaplan, 1965). Such single mutations markedly affecting a character rarely occur in animals, so plant domestication is probably more easily and rapidly accomplished than animal domestication.
   Most of the easily selected characters, that is, ones alterable by mutations in one or a few genes, were probably already incorporated into crop plants in prehistoric times. However, the facility of incorporation of the mutant gene would depend on whether it was dominant or recessive and on the plant's breeding system, being easier to manipulate in

inbreeding plants than in outbreeding ones. But, as Gustafsson (1947) has written, single mutations of economic value must now be regarded as of little importance in plant breeding programmes. Consequently, plant breeding by man comprising deliberate crossing and selection has been concerned, as in animal breeding, primarily with quantitative characters.

The practice of plant and animal breeding assumes that artificial selection increases the frequency of favourable alleles over many loci, with consequential advantageous effects on quantitative characters. The corresponding genetical theory is that, for most characters, a population contains 'multiple alleles at each of numerous loci distributed over each of many paired chromosomes which may express average, dominant and epistatic gene effects in conjunction with random environmental effects on the population for any measure of phenotype' (Dickerson and Willham, 1983). Actually, as demonstrated in Nicotiana (Jinks, 1981) and in maize (Sprague, 1983), it is probable that most genetic effects in hybridisation are exerted through dominant allelic interactions; overdominance (interallelic interactions producing unique hybrid substances) and epistasis (nonallelic interactions where one locus influences the expression of another) do occur, but only rarely. There seems to be no reason to believe that heterosis (hybrid vigour) is 'unfixable', so it should be possible to establish pure-breeding lines which are at least as vigorous as the hybrid generation.

Certain adaptations, especially physiological ones, are less sensitive to single gene inactivation; these include domestically undesirable features such as seed dormancy, perennial habit, indeterminate growth and daylength sensitivity. They also probably became modified in prehistorical domestication and breeding, but more slowly, because such characters are often determined by many genes with cumulative effects. Unconscious selections by man probably did not increase yield very much but they did ensure uniformity of growth and reproduction and an improvement in cultivability and harvesting. Zohary (1969) reports that the yields of wheat and barley in the Middle Ages were no higher than those of wild wheat (Triticum dicoccoides) and wild barley (Hordeum spontaneum). However, since then, average wheat yields have risen from between one-half and three-quarters of a tonne per hectare to about 5.4 tonnes per hectare in England in 1980 (Evans, 1980).

Deliberate selection by man has been mainly concerned with improving seed, fruit or nut yield, oil, starch or sugar content, drought, cold and disease resistance, and flavour and baking quality, all multigenetically determined. Modern plant breeding and agronomy have been particularly successful in increasing yield. The accelerating grain yield of wheat in England, and the estimated proportion of yield increase that is due to genetic influence, are shown in Figure 10. Yields have risen steadily in wheat since about 1940 and at a markedly greater rate than in previous times. There is no sign that the rate of increase is slowing down, but the physiological means whereby genetic changes have caused higher yields are rather unclear. Since primitive varieties of cereals may outyield modern ones under poor conditions, Evans (1980) has concluded that selection has been for greater responsiveness to the cultural practices of man such as the use of fertilisers, irrigation, fungicides and insecticides, weed control etc. This has resulted in wheat in increased grain size, mostly due to an increase in endosperm cell number, and in the number of grains per spikelet (Dunstone and Evans, 1974).

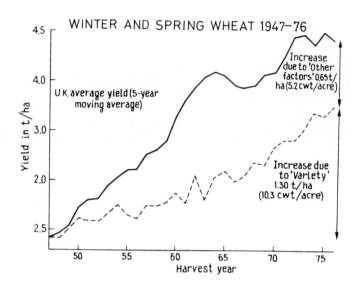

Figure 10 Five-year moving averages for the yield of wheat in the United Kingdom from 1947 to 1976. Solid line, total yield; dotted line, yield due to 'variety' (Riley, 1980).
The genetic component of the yield increase is about 60%, the remainder being due to improved agronomy.

As compared with wild species, rates of photosynthesis were found to be unchanged in maize, sugarcane, sugarbeet and sorghum, while in wheat and cotton it has apparently fallen during domestication (Evans, 1976). Also, relative growth rates have not increased during domestication in wheat, maize, tomatoes and cowpeas. The main physiological character that has improved with breeding is the cross-sectional area of phloem in the culm of wheat and oats, which allows an increased capacity to transport sucrose to the larger developing grains (Evans et al., 1970).

In raising the potential yield of crops, there has apparently been a progressive increase in the proportion of the plant that has been partitioned into the harvested organs. The total mass of the highly bred plant has not altered from its progenitors, but the distribution between vegetative and reproductive organs has changed. This proportion increased from 5 to 17% in wild diploid wheat to about 51% in a modern variety (Austin et al., 1982), and similar increases have been found in barley, soybeans and peanuts. According to Evans (1983), this has been achieved by a genetic reduction in stem length, a longer period of phtosynthetic activity in the upper leaves, a greater duration of grain growth and an enlargement of the phloem.

The practice of artificial selection in plants has evidently produced pronounced alterations in morphological and physiological phenotypes. While some of the quantitative changes are due to improved cultivation procedures, many others and most qualitative ones are the result of modification to the genome. We are particularly interested in the means whereby these genotypic changes are expressed, because of their evolutionary implications. According to Wareing (1978), it is commonly considered that the use of growth regulators (hormones) and breeding are alternative ways of achieving the same ends. Either the application of exogenous hormones or a breeding programme has been shown to enhance growth in many vegetables, increase sugar content in sugarcane, improve drought resistance in cereals, increase apical dominance etc. (Peacock, 1976). Application of gibberellic acid to dwarf plants converts them into tall ones, while inhibitors of gibberellic acid turn tall plants into dwarf ones (Cleland, 1974). Moreover, single gene mutations affecting hormone synthesis, metabolism or sensitivity are known to produce pronounced phenotypic effects, and some of these mutations have been incorporated into cultivars. The synthesis of hormone is affected in a diageotropic mutant of the tomato, which has a reduced level of ethylene (Zobel, 1973), while the synthesis of gibberellic acid is elevated in the mutant, *gigas*, of barley (Favret et al., 1975). Other hormone mutants, like the Norin 10 semidominant dwarfing genes in wheat, confer insensitivity to gibberellic acid (Gale and Law, 1977) and the viviparous embryo mutants of maize have reduced sensitivity to abscisic acid (Robichaud et al., 1980). Frequently, mutants in plants are a consequence of alterations in the metabolism of hormones. A red clover mutant described by Stoddart (1966) appears to be unable to flower because the mutation leads to a rapid degradation of gibberellic acid under long-day conditions. An abnormal interaction between gibberellic acid and abscisic acid has been reported in strains of peanuts with a prostrate (runner) instead of an erect habit (Halevy et al., 1969). The prostrate strain, which could be converted to the erect form by the application of gibberellic acid, contained an inhibitor of gibberellic acid with the characteristics of abscisic acid.

Plant hormones, or plant growth regulators, are like animal hormones in that they are synthesised in one tissue, move about the plant and are active in coordinating growth at very low concentrations in tissues distant from the source. They are unlike animal hormones in being less specific in their effects, often acting synergistically with other growth substances, and their activity may be much less concentration dependent. Nevertheless the evidence from single-gene mutants shows that hormones are involved in the specification of plant height, growth habit, flowering, sex expression, parthenocarpy and seed dormancy. Corresponding mutations are indicative of genetic control which, taking into account hormone synthesis, metabolism and transport, the presence of inhibitors, hormonal interactions and the affinities and position of receptors, would constitute a multigenic hormonal determination of a particular economic character. The main targets of selection in plant breeding are probably genetic variations effecting slight alterations in hormonal levels and interactions leading to small changes in phenotype; e.g. different patterns of indole acetic acid and abscisic acid in response to drought stress in sorghum (Simpson et al., 1978).

In other respects, according to Pickersgill and Heiser (1976), crop plants have remained much the same genetically since they were first dosmesticated. Alterations in nuclear DNA by polyploidy or repeated sequence amplification do not seem to be involved in the attainment of breeding objectives. Furthermore, there is no evidence that multiple copies of unique genes arising from polyploidy or sequential duplication in crop plants have diverged significantly in function. Even so, the evolutionary distance between wild and domesticated species is believed to be greater in plants than in animals, probably because stronger selection pressures have been used in plants, larger populations have been subjected to selection, and the relatively less complicated nature of plant development has allowed the isolation of more radical mutants.

Animal domestication and breeding
A prerequisite to breeding is the domestication of the animal to make it docile, so that controlled crossing and selection may be undertaken. The transfer of animals from the wild to the domesticated condition is not usually a simple matter of keeping them in captivity until they become tame. Even when such tamable animals as deer and foxes are so domesticated, their offspring born in captivity have a wild form of behaviour (Hagedoorn, 1944). This is because there has been no modification of the genotypes of the parents, so the young are still instinctively taught the forms of behaviour appropriate to natural environments. For genetically transmissible docile behaviour, it is usually necessary to select in successive generations for increased docility in young animals kept apart from the parents. This has been done with silver foxes by Belyaev (1980), where tameness was found to depend on appropriate alleles at many loci. The alleles appear to be of loci concerned with the levels or interaction of reproductive hormones which, in the wild, are stimulated by daylength to initiate seasonal sexual reproduction.

A correlate of selection for docile behaviour is a loss of the strict seasonal pattern of reproduction in wild mammals, with a consequent increase in fertility. In foxes selected for domestic behaviour this is accompanied by changes in the neuro-endocrine system. Alterations occur in the levels of corticosteroids produced by the adrenal gland, of steroid sex hormones and of the regulatory hormone serotonin. Selection for domestication in many animals is apparently achieved by a genetic rearrangement of both the central and peripheral hormonal control of ontogeny, which Belyaev (1980) interprets as providing an increased resistance to stress, including the presence of man himself.

Some other groups of mammals like rabbits and rats are untamable, meaning that insufficient natural variation in behaviour exists for selection for tamability to be successful. However, in some of these animals interspecific or interracial crossing generates sufficient behavioural variation in the segregating offspring for selection to be effective. For example, the wild rabbit is not tamable but the hare is, and hybrids between the two are readily tamable and fertile. In fact, the breeding of tame rabbits in France can be traced back to the production of such 'leporids', from which originated the domestic breeds of rabbits (Hagedoorn, 1944).

The order of animal domestication as inferred from archaeological evidence is roughly in accord with the degree of tamability of the species as determined in attempts at domestication. Thus sheep, goats and reindeer were followed by cattle and horses, and still later by less tractable animals like cats and polecats (Zeuner, 1963). Selective breeding within these domesticates has usually resulted in separate breeds, each of which contains a number of semi-isolated inbred lines, e.g. about 800 existing recognised sheep breeds and more than 200 breeds of domestic goats (Spillett et al., 1975). In the process, there has been a large decrease in genetic variation from the wild condition in any one line, but some degree of diversity is preserved as genetic variability between lines. The necessary inbreeding must have involved a drastic change in the animals' genetic constitution; modern experiments show that in the house mouse only one line out of 20 survived 12 generations of inbreeding (Bowman and Falconer, 1960), in *Drosophila* 10% of lines survived for more than 20 generations (Clayton et al., 1957), and in the fowl a mere eight out of 279 lines remained following prolonged inbreeding (Abplanalp, 1974). Most natural populations of animals consist of highly heterozygous and heterogeneous individuals, and inbreeding facilitates the expression of numerous lethal, subvital and other recessive alleles. A consequence of the establishment of viable inbred lines is that heterosis (hybrid vigour) is weaker than in many plant hybrids and often decays after the first hybrid generation (Sheridan, 1981). A more serious consequence is that inbred lines may become hosts for disease as the result of the fixation of one or a few genes. Rendel (1984) cites examples in mice where such homozygosity predisposes different lines to infection by flaviviruses, or orthomyxoviruses and by lactate dehydrogenase elevating virus when the animals are immunologically suppressed.

In the actual breeding of animals, single-gene mutations are rarely of importance, unlike the situation in plants, presumably because the closely knit action of most developmental genes in animal ontogeny means that a drastic change in one affects many other processes. A few mutations of major genes are currently used in animal breeding, such as dwarfing genes in fowls, genes related to meat quality in pigs, a gene for multiple ovulation in sheep and muscular hypertrophy genes in cattle (Cunningham, 1984). But most animal breeding involves the selection of naturally occurring alleles of small effect, with many alleles determining an economic character; e.g.

the important wool protein, keratin, is coded for by at least 50 to 100 genes (Cunningham, 1984). In essence, animal breeding consists in the substitution of one allele by another, and the selective response predicted for a character is determined by the heritability (the proportion of the total genetic variance that is due to additive gene action).

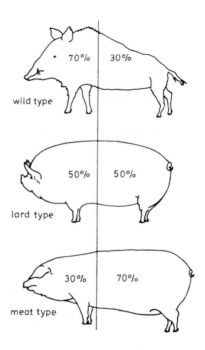

Figure 11 Diagram to illustrate changes in the proportion of body parts in pigs that accompany domestication and breeding (Borojevic, 1980).

Objectives in animal breeding include increase in milk production in dairy cattle, muscular mass and distribution in pigs (Figure 11), wool production in sheep, number of eggs in fowls etc. The breeding practices have been successful to such an extent that whereas wild cattle gave only a few hundred millilitres of milk, the best bred individuals yield up to 15 000 litres during the lactation period, wool production in sheep has risen from one kilogram to more than 20 kilograms per animal, and conversion of food into meat in pigs has improved in efficiency from 4.5 to 3.5 pounds of grain per pound body weight in 50 years of selection. Such pronounced responses to selection are due to the high degree of genetic variation exhibited, particularly by different breeds or lines of domesticated animals. In selection for beef production, for example, while breeds show few genetic differences in muscle or bone distribution, there is much variation in speed of maturity, proportion, partitioning and distribution of fat, and in muscular weight relative to live weight, all characters of economic importance and targets of selective breeding (Berg and Walters, 1983).

As far as can be determined, many of the genetic advances in animal breeding result from alterations in hormone levels due to differences in synthesis or metabolism and to changes in the number, positions and affinities of hormone receptors. Metabolic hormones affecting growth and development, probably interacting with each other and with growth factors and usually having multiple sites of action, include insulin, growth hormone, glucagon, thyroid hormones, adrenal cortical hormones, catecholamines, somatomedins and anabolic steroids. A knowledge of the relation of these hormones to changes in economically important characters is probably most advanced in pigs. Thus, as compared with ordinary pigs, lean ones have higher concentrations of plasma growth hormone and somatomedins and a higher resistance to insulin, but no change in the level of plasma insulin (Wangsness et al., 1977). The growth hormone stimulates protein synthesis and the growth of muscle, while the somatomedins (peptides produced by the liver etc.) increase cell division. Such pigs also possess a reduced level of adrenocorticotrophic hormone, which, while apparently stimulating meat production, renders the animals more susceptible to stress (Schönmuth, 1980). Direct selection for hormone activity has been used to eliminate stress-sensitive pigs (Smith and Brampton, 1977) and poultry (Brown and Nestor, 1974), and to increase the reproductive rate of sheep (Land and Carr, 1979).

The use of applied hormones in meat, milk and egg production, as in reproductive behaviour, supports the generalisation that genetic changes in these characters are exerted through hormonal change. Thus exogenous growth hormone produces anabolic responses in cattle, sheep and pigs (Moseley et al., 1982). In dairy cattle, the injection of bovine somatotrophin increases milk production by up to five litres per day (Harvey, 1986), apparently by mobilising body fat and diverting glucose and fatty acids away from the formation of tissue. Protein gain in cattle is positively correlated with concentration of plasma insulin (Eversole et al., 1981) which increases the intracellular supply of amino acids and the activity of initiating factors in protein synthesis. In rats, insulin plays a dominant role in fatty-acid synthesis and in the accumulation of lipids in adipose tissue (King and Mainwaring, 1974).

The evidence in animals for many of the results of domestication and breeding practices being brought about by alterations in hormonal systems is even stronger than it is in plants. Artificial selection is also likely to involve changes in metabolism and perhaps in other processes which, to the extent that they are not influenced by hormones, depend on different sorts of genetic variation.

Hormonal change and artificial selection

Hormones act to regulate pre-existing processes in multicellular organisms. They may influence the rate of enzyme synthesis, affect the rate of enzymatic catalysis or alter the permeability of cell membranes. The activity of some hormones like vasopressin is directed towards the rapid modulation of the enzymatic or transport activity of proteins, other hormones like ecdysone are directed towards the control of gene expression, and some like insulin carry out both functions.

A characteristic of hormonal effects is that they are often balanced by the counteracting effects of other hormones; e.g. both glucagon and epinephrine promote the release of glucose into the blood stream from liver glycogen. Also, hormone action is usually part of a feedback loop; e.g. the uptake of glucose by tissues is promoted by insulin, but the lowered glucose concentration of the blood leads, through feedback inhibition, to a decreased rate of secretion of insulin from the pancreas. The net result of these connected reactions is a relative constancy in the concentration of important metabolites in the tissue fluids. Thus, in the control of blood sugar level, a balance is maintained among the processes that cause the breakdown of glucose in the tissues (glycolysis), the conversion of glucose to glycogen (glycogenesis), the conversion of liver glycogen to blood glucose (glycogenolysis) and the formation of glucose from noncarbohydrate precursors (gluconeogenesis).

A change in the concentration of one hormone will thus affect the concentration of many other hormones and the processes they regulate. Consequently, the buffering or homeostasis of the organism that gives resistance to perturbations induced by genetic or environmental factors will be weakened. In particular, previously unexpressed genetic variation will now be exposed, as has been demonstrated by Waddington (1953) and Milkman (1961). It may often be the case that breeding advance is by the utilisation of the genetic variation revealed when the hormonal balance of the organism is upset. Belyaev (1980) has observed that selection for tameness in silver foxes results not only in a change in the level of steroid sex hormones, but also in a much higher frequency of appearance of variants affecting tail and ear morphology and coat colour.

A breakdown in hormonally induced homeostasis may also be involved in the natural process of species formation. Speciation is said to require a drastic reorganisation of gene pools (Mayr, 1963) or a disruption of closed variation (Carson, 1975). These proposals refer back to the belief that selection does not act upon individual mutations but upon interacting complexes of genes. The argument is that the adaptive value of a gene product depends not only on the external environment of the organism containing it but also on many other gene products within the cell with which it interacts. In this model, the process of species formation is thought to involve a shift from a well-balanced gene pool, through an unstable genetic constitution, to a new integrated and balanced population.

2.1.3 *Conclusion*

A brief examination of the development of genetic resistance to toxins indicates that, at least over short periods, only relatively slight adjustments can be made in metabolic systems. They are usually confined to decreases in permeability, alterations in binding sites, amplifications of resistance genes and changes in the specificity of pre-existing enzymes. New genes coding for new enzymes that can inactivate a toxin by the hydrolysis of a bond or by replacement of a chemical group have not yet appeared. When we consider the extreme improbability of formation of even a relatively small gene by random polymerisation, it cannot be expected that new genes for detoxifying enzymes will arise *de novo*. Resistance genes already present must have been gradually built up from a short sequence specifying a small catalytic polypeptide or by 'recombinational' mechanisms involving the rearrangement of pieces of previously evolved genes. Only when there is already an enzyme with a certain level of affinity for the toxin, can point mutations increase its affinity to detoxify the compound. The potential for such changes in specificity therefore depends on the presence of previously evolved appropriate enzymes, and seems to be lower in fungi and plants and higher in bacteria and insects.

In contrast, evidence from plant and animal breeding strongly suggests, especially in animals, that evolution by artificial selection is largely exerted through alterations in hormonal systems. Many of the altered characters appear to be due to changes in developmental processes affecting tissue distribution, organ size, interactions with environment and changes in behaviour.

The hormonal system of animals and plants consists of the producing gland or tissue, the hormone itself, its rate of synthesis, secretion and transport, its metabolism, the presence of inhibitors and activators, the induction and structure of its receptor and often its binding sites in the nuclear DNA. While, in theory, genetic alterations could affect any part of the hormonal system, changes in some components will be more specific in effect or lead to larger alterations in phenotype than changes in other components. At the subspecific level, artificial selection commonly results in quantitative changes in the levels of circulating hormones. Examples have been given of alterations in rates of synthesis in barley and pigs, increases in rates of degradation in red clover, formation of hormone inhibitors in peanuts and hypertrophy of the producing gland in silver foxes. A permanent alteration in hormonal level affects the regulatory network and relaxes homeostatis to expose hidden genetic variation; both may be important in breeding practice.

As we shall see in the next Section (2.2), an important mechanism in the formation of lower taxa like species and genera is the alteration of the specificity of gene expression. This occurs by changes in the sites of binding of hormones and other factors that flank the coding regions of most genes.

The evolution of higher taxa like orders and classes sometimes involves alterations in the hormones themselves. Thus the polypeptide hormones secreted by the pancreas, pituitary and hypothalamus of animals can be changed slowly at the level of class because they are direct products of translation. For example, complete or partial gene duplications, followed by single base substitution, have generated a variety of neurohypophyseal hormones in vertebrates (Figure 12). These include arginine vasotocin (in nonmammalian vertebrates), arginine vasopressin (in most mammals), lysine vasopressin (in pigs) and isotocin (in teleost fish), each of which differently affects uterine contraction, blood pressure, diuresis, milk ejection etc. when tested in the rat or rabbit (Wallis, 1975). On the other hand, any change in structure of a nonpeptide hormone would require the origination of a new enzyme to modify it, or an alteration in the specificity of an existing enzyme. Since they rarely if ever occur, the steroid hormones of the lower vertebrates have an adrenocortical pattern of secretion closely resembling that of the mammals; i.e. either corticosterone or cortisol, or both, constitute the major corticosteroid output in species of fish, amphibia, reptiles and birds, just as they do in mammals (Barrington, 1964).

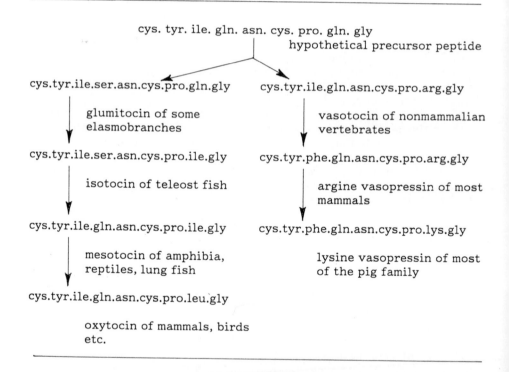

cys. tyr. ile. gln. asn. cys. pro. gln. gly
hypothetical precursor peptide

cys.tyr.ile.ser.asn.cys.pro.gln.gly

glumitocin of some elasmobranches

cys.tyr.ile.ser.asn.cys.pro.ile.gly

isotocin of teleost fish

cys.tyr.ile.gln.asn.cys.pro.ile.gly

mesotocin of amphibia, reptiles, lung fish

cys.tyr.ile.gln.asn.cys.pro.leu.gly

oxytocin of mammals, birds etc.

cys.tyr.ile.gln.asn.cys.pro.arg.gly

vasotocin of nonmammalian vertebrates

cys.tyr.phe.gln.asn.cys.pro.arg.gly

argine vasopressin of most mammals

cys.tyr.phe.gln.asn.cys.pro.lys.gly

lysine vasopressin of most of the pig family

Figure 12 One scheme for the evolution of the neurophypophyseal hormones of vertebrates. A duplication of the original gene is indicated by the presence in primitive fish of both vasotocin and oxytocin. Alterations in hormonal function were subsequently brought about by simple evolution involving amino-acid substitutions (data from Wallis, 1975).

For both the peptide and nonpeptide classes of hormone, the structure of their receptors appears to have been highly conserved during evolution. Muggeo et al. (1979) showed that the affinity of receptor for insulin has remained constant in all vertebrate species from bony fishes to man and that the hormone-binding part of the insulin receptor has been even more highly conserved than the hormone itself. This constancy is believed to reflect a co-evolution of peptide hormones and their receptors over 400 million years.

Alterations in hormonal systems may be able to account for many advances in plant and animal breeding, as well as being a factor in the formation of species and genera, but they do not seem to be involved in the formation of higher taxa.

2.2 *Evolution at the Species Level*

In the evolution of resistance to toxic agents, selection pressure has been intense but only over a few hundred generations. In such circumstances, resistance to the poison takes precedence, often at the expense of general adaptation. Similarly in breeding, the aim is to produce hypertrophies of various organs and the dismantling of ecotypic adaptations for the benefit of man. As shown by the inability of many highly-bred organisms to maintain themselves in the wild, human objectives are often achieved at the expense of general adaptation. Therefore, Darwin (1859) and Haldane (1954) were correct in their views on the efficiency of artificial selection but perhaps not on its relevance to the course of evolution. Nevertheless, breeding for resistance to toxins and for economic ends illustrates how mutations in enzyme activity and in hormonal level or activity can contribute to evolution, at least in the short term.

The formation of a new species takes about three million years on average (Table 4, p. 36), selection pressures are usually quite low, and the organism is thought to remain in adaptive balance with its environment throughout the process. It has been repeatedly observed that, as judged from electrophoretic variation in proteins, due to amino acid substitutions causing changes in net charge, populations become genetically more different from one another as they diverge reproductively and morphologically. But such observations do not disclose whether the measured change in variation with divergence represents, or is connected with, the genetic changes responsible for species formation. It is possible that, following reproductive isolation, certain mutations steadily increase with time because they are insensitive to natural selection.

To examine this matter, Avise and Ayala (1975) compared electrophoretic variation at 24 loci in two groups of fish which differed 25-fold in the number of living species. The comparisons were made on the reasonable assumption that, if electrophoretic differences were symptomatic of the genetic changes involved in speciation, a group of organisms with the greater number of species would show the greater interspecific divergence.

The data were expressed as genetic distance, a statistic derived from the logarithm of the value for genetic identity; it is a meassure of the average number of detectable substitutions per gene that have accumulated since the populations became separated from a common ancestor. The average genetic distance among nine genera of minnows of the family Cyprinidae, which has about 250 species, was 0.568. This value is slightly lower than 0.627 for the sunfish (*Lepomis*), which has only eleven species altogether. The degree of genetic difference in the two groups does not support the idea that point mutation, at least in genes making enzymatic proteins, is related to species-forming processes; rather, it seems to be more a function of the time that the species have been separated. Another investigation with the same conclusion is that of Wilson (1976) who compared frogs having thousands of similar species with placental mammals, divided into about 20 morphologically diverse orders: the rate of accumulation of point mutations is not correlated with morphological change.

2.2.1 *Apparent noncorrelation between protein and morphological evolution*

If the problem is approached from a different angle, the matching of protein change to specific morphological difference, the conclusion again is that there is little correlation between the two.

The primate species, man and chimpanzee, belong to different families, which diverged five to ten million years ago. They differ markedly from each other in a number of anatomical features such as locomotory, masticatory and cranial characters. Nevertheless, the amino acid sequences of 12 intracellular proteins were found to have greater than 99% identity in the two species, as also had over 30 other proteins when tested by electrophoresis (King and Wilson, 1975). Although the two primates are morphologically distinct, the protein data place them as close genetically as sibling (morphologically similar) species of other mammals and of *Drosophila*.

Investigations with essentially similar results have been made with species of Lepidoptera (Brittnacher et al., 1978), with salamanders (Larson and Highton, 1978) and with fishes (Kirkpatrick and Selander, 1979). All show that there is little correlation between rates of enzymatic sequence change and rates of morphological diversification.

King and Wilson (1975) have therefore proposed that most evolutionary divergence results from changes in development which are reflected as altered morphology. However, in the primate example, the proteins examined represent less than 0.1% of each genome's coding capacity, and such proteins are unlikely to play a part in determining any of the morphological differences between the families. Moreover, if most protein changes are neutral to natural selection, the few that might be developmentally significant would be masked and therefore undetectable.

2.2.2 *Evidence for developmental evolution*

Adaptation occurs by both metabolic and developmental change, but it is generally believed that the former, having originated very long ago, is rarely responsible now for evolutionary advance. Consider the ornithine cycle, which provides one of the most complete and informative examples of metabolic evolution. It seems first to have evolved in bacteria for the synthesis of arginine, and later in land vertebrates it became a means for the detoxification of ammonia. However, since its early Precambrian origin, the only form of evolution appears to be that of regulatory adjustment.

In some bacteria, the breakdown of arginine in the presence in the cell of arginine synthesis has necessitated a gene duplication to specify two separately regulated enzymes (Paulus and Switzer, 1979). The paths of arginine metabolism have remained much the same in the fungi but the first steps in arginine formation take place in the mitochondria, while arginine catabolism is confined to the cytosol (remaining cytoplasm) (Jauniaux et al., 1978). Another dimension to regulation, that of intracellular compartmentation, was added with the evolution of the organelle-containing eukaryotic cell.

The disposal by marine predators of excess nitrogen is by secretion into the water of ammonia, but this is not possible in land animals with a low water intake. Consequently, the ornithine cycle has been used by most amphibia, some reptiles and all the mammals to convert ammonia to urea, which is both nontoxic and water soluble. This has apparently been done by a slight change in the regulation of one enzyme, a reduction in the levels of other enzymes, an increase in the permeability of mitochondria to ornithine and a change in allosteric inhibition to increase urea synthesis (Paulus, 1983). Nearly all the adaptive changes have been regulatory ones concerned with nitrogen economy or with the efficiency of nitrogen waste disposal. It appears that no entirely new substrate specificities or types of bond rearrangement have evolved in any of the enzymes over at least 1000 million years.

In contrast, morphological evolution of roughly comparable adaptive significance can be at least 1000 times more rapid. We can take a typical example from the fishes. Small freshwater fishes of the family Cichlidae have undergone extensive speciation since the mid-Pleistocene (about 800 000 years ago) in the lakes of east and central Africa. Lake Victoria, for example, contains an estimated 165 species of cichlid fishes which have become individually specialised for diverse food sources (algae, phytoplankton, higher plants, zooplankton, insects, whole fishes and fish scales), thus occupying virtually all ecological niches in the lake, despite strong competition from fishes of other families (Greenwood, 1973). They are enabled to do this by the gradual evolution of a new adaptive complex involving slight but significant shifts in positions, proportions and fusions of bones and muscles affecting the jaw and its articulation (Liem, 1974). Added to this altered jaw structure were modifications of the teeth, gill rakers and gut, to enable specialisation in particular foods. As far as can be seen, both the evolutionary novelty and the subsequent adaptive modifications were developmental (morphological), but it can be presumed that there were accompanying biochemical adjustments, especially in digestive enzymes.

2.2.3 *Regulatory change in developmental evolution*

The genes directly responsible for processes of differentiation and morphogenesis would be expected to be genetically modifiable only with difficulty because they are usually pleiotropic in their activites. Pleiotropy in development may be relational where there is a single site of gene action on which all other effects are consequent, or direct where diverse effects are caused by the direct action of a single gene (Grüneberg, 1963). A mutation will result in relational pleiotropy, for example, when it inactivates or otherwise affects a component of a hormone system (producing gland, transport protein, hormone or receptor structure). Such mutations occur often in mammals and probably in plants, where they affect the expression of multiple genes. Direct pleiotropy will occur when a mutation affects a gene that is expressed in a number of different tissues or organs or at different stages in the life cycle. An example in *Drosophila* is a mutation affecting a membrane component involved in interactions between cells (Poodry et al., 1973). The mutant gene is detectable at six critical periods in development, and causes paralysis, scarred eye facets and altered bristles on the head and thorax. This type of pleiotropy is infrequent in flowering plants, which lack cellular movements and inductive interactions in development.

Although pleiotropy ensures that alterations in basic developmental processes are usually inviable, these processes are controlled or regulated by numerous other genes which are much more readily altered. This has been shown by the occurrence of disturbances in ontogeny in hybrids between species of fishes that exhibit nearly identical morphologies. These disturbances are reflected in increases in nonhatching of eggs, in the frequency of morphological abnormalities and in earlier developmental arrest and lethality (Parker et al., 1985). The genetic differences between the species must be in the regulation of morphology and not in the basic processes, because the morphologies of the different species are very similar.

The nature of the regulatory changes was determined by alterations in enzyme activities and in the expression of isozymes (variant forms of the one enzyme). Generally, the changes observed have been specific to particular enzymes and tissues, and do not affect many enzymes of the same tissue or the same enzyme in many tissues (Kettler et al., 1986). Quite frequently in hybrids, there are delays in the time of first appearance of an enzyme which is believed to be due to slight incompatibilities between effector molecules (*trans*-acting factors) and gene regulatory sites. With regard to the enzymes that are specific to tissues, hybrids between closely related species may express the corresponding genes, but at different developmental times and at different levels of activity. In hybrids between more distant species, the tissue-specific isozymes may not be expressed. Often in hybrids, the paternal form of a given gene (allele) for a particular enzyme is more retarded than the maternal allele in its expression. Ohno (1969) has attributed this phenomenon to the regulatory sites of the paternal genome not interacting with the maternal gene regulatory molecules carried over from oogenesis or synthesised by the maternal genome during embryogenesis.

2.2.4 *Structural and regulatory genes*

A distinction is often made between structural genes and regulatory genes. The majority of structural genes encode catalytic proteins (enzymes) and noncatalytic proteins such as membrane subunits and viral coats. Some structural genes like those for transfer and ribosomal RNAs do not encode proteins and not all protein-coding genes can be called structural because ones that encode receptor proteins, polypeptide hormones and DNA binding proteins are part of the regulatory class. Yet other genes specify proteins that contain a catalytic and separate regulating site in the one molecule. Regulating genes are ones that determine or influence the timing, placement or degree of structural gene action. Even such a functional definition of regulatory genes is not absolute, for there are proteins which have both a structural (catalytic) and a regulatory role. Although there are genes, such as those making hormones, repressors and activators whose primary role is to specify gene expression according to stage of development or type of tissue, many others have regulatory sequences in or near their transcription initiating regions (promoters). The DNA sequences of these genes thus include both an enzyme-specifying and a regulatory component. In view of these complications, it seems best to speak not of structural and regulatory genes, but of structural and regulatory genetic changes.

Regulatory processes occur in enormous variety at many or all levels of gene organisation and function including transcription, messenger RNA processing and transport, translation and activity. However, Zuckerkandl (1978) has quoted eight investigations which show that translational control is not generally a predominant factor in gene expression. Nor does control through the processing of RNA transcripts appear to be common, although it apparently occurs during sea urchin development (Kleene and Humphreys, 1977). Transcriptional changes in gene expression, on the other hand, markedly and commonly affect morphology, as shown by such phenomena as metamorphosis, caste specification and sexual dimorphism. All these result from the programmed switching on and off of gene transcription.

2.2.5 *Frequency of regulatory versus structural change*

One way of detecting regulatory gene divergence among closely related species is to examine differences in levels of expression and developmental patterns of enzymes in polyploids of ancient origin. Ferris and Whitt (1979) examined the expression of duplicate genes in ten tissues from each of 15 species of catostomid fishes 50 million years after the polyploidisation event that gave rise to the family. About half of the duplicate genes had lost expression during that time, but the expression of those remaining had altered in nearly 60% of the tissues examined. Duplicate gene expression seems to have changed mainly in respect to different tissues and organs and not in regard to cell types within a tissue or to stages in development. Of 84 divergent tissue patterns of gene expression, only 12 were believed to be solely explained by mutational changes in the enzyme-coding regions of the gene. It was concluded that most, if not all, of the tissue-characteristic enzyme patterns were brought about at the level of enzyme synthesis and probably at the level of transcription or processing of the nuclear RNA transcripts.

The evolutionary significance of such divergences in gene expression may, however, be affected by the possibility that the duplicate loci of polyploids facilitate regulatory changes. Nevertheless, Crawford (1975) has commented, with regard to an evolutionary survey of the tryptophan biosynthetic pathway in nonpolyploid bacteria, algae and fungi, that 'the regulatory mechanisms can be altered during evolution at least as easily as chromosomal gene location and much more easily than the amino acid sequence of an enzymatically active polypeptide.'

2.2.6 *Regulatory changes in developmental genes*

In the study of gene expression in hybrid fishes (Kettler et al., 1986), an altered enzyme appearance was correlated in a few instances with an alteration in tissue differentiation, suggesting that the latter was the primary effect of the genetic change. However, in many other instances, a change in the timing of gene expression was independent of morphogenic events. In *Dictyostelium*, for example, most developmentally regulated functions have been shown to be nonessential for development (Kaiser, 1986). Evidently the majority of genes that are expressed in a particular structure have nothing to do with its formation but are developmentally regulated genes concerned with its metabolism or maintenance, or are merely structurally partitioned in expression. The existence and number of morphogenic genes can be approximately established by mutational analysis of a structure; thus, in the developmental formation of conidia in *Asperigillus*, it is estimated that 45 to 150 genes are involved (Martinelli and Clutterbuck, 1971). However, the measurement of messenger RNA diversity based on nucleic acid hybridisation indicates that about 5000 genes are expressed in conidial cells of the closely related genus, *Neurospora* (Dutta and Chaudhuri, 1975). The genes responsible for morphogenesis are likely to include those specifying tubulins, actins, myosins and cell surface proteins involved in cell shapes and movements, but are only a few of those expressed in a structure.

Edelman (1983) considers that form in insects and vertebrates, but not apparently in annelids and plants, arises as a result of the differential effects of the driving forces provided by the processes of cell division, cell movement and cell death. The regulation of movement and position is carried out by particular mechanisms of cellular adhesion. For example, before induction, all nonmoving cells in the embryonic tissue (mesenchyme) that develops into adult connective tissue etc. have both neural and liver cell adhesive molecules at their cell surfaces. At neural induction, a central region of the blastoderm, the incipient neural plate that gives rise to the central nervous system, increases the neural, and decreases the liver, adhesion molecules. This reciprocal change, which persists through life, must occur by alterations in gene expression. The cells in the band forming the neural crest lose their neural adhesion molecules on the surface when they migrate to various sites. When in position, the neural adhesion molecules reappear on the surface, and the cells aggregate and differentiate to form the ganglia. The neural cell adhesion molecules of a cartilaginous fish or a frog bind to those of a mouse as readily as to themselves (Hoffman et al., 1984), indicating that it is not the adhesion molecule itself that changes but rather the regulation of the time and place of its expression.

A relatively simple example of evolutionary changes in morphogenic genes is that of the chorion proteins that make up the shell of the silkworm egg (Kafatos et al., 1977). The main features of the surface morphology of the egg are the predominantly lamellar pavement substructure with internal vaulting of the eggshell and the chimney-like respiratory structures called aeropyles on its surface. These structures are made by more than 100 proteins belonging to several related families encoded by three clusters of genes on a single chromososme of *Bombyx mori*. They are expressed in a time-dependent and cell-specific manner. In two species of silkworm, *Antheraea polyphemus* and *A. pernyi*, there is a pronounced difference in the aeropyle on the eggshell surface. Although there have been some changes in chorion structural genes, the morphological differences are primarily due to changes in gene expression. The relative timing of expression of chorion genes has not altered, the specific differences rather being due to a quantitative change in gene expression giving different amount of filler (substance moulding the aeropyle lamellae into their particular shape).

An intergeneric difference between *Antheraea polyphemus*, which has aeropyles, and *Hyalophora cecropia*, which lacks them, also results from regulatory changes in the expression of chorion genes that make the filler substance (Hatzopoulos and Regier, 1987). The failure to form aeropyles in *Hyalophora* is correlated with a reduction in the amount and number of proteins synthesised in the very late period of eggshell formation, probably affecting both lamella-forming protein and one of the proteins that assemble to make the filler substructure. Again, only minor changes in the timing of gene expression were found, suggesting that the control elements for timing are separate from those that control amount and cell specificity.

These two examples of changes in genes affecting insect eggshell structure suggest that the evolutionary alteration of morphology is not very different from developmental change in enzyme activity. Like the genes for chorion proteins, many developmentally-regulated structural genes such as those for actins, keratins and tubulins are organised into multigene families, the members of which have been derived by successive duplications of an ancestral gene. Such an organisation, by providing multiple genes with different temporal and spatial expressions, is believed to allow evolutionary flexibility by enabling a change in one member without affecting the expression of others.

2.2.7 *Adaptive significance of regulatory change*
Despite the claims that changes in gene regulation may be more important in adaptation than changes in protein-coding sequences, most of the supporting evidence is of an indirect nature in eukaryotes. In prokaryotes,

on the other hand, there are numerous examples of the utilisation of novel carbon sources by genetic alterations of the inductive system (Ornston and Parke, 1977; Clarke, 1980). In most instances, enzyme synthesis is rendered constitutive (continuously produced) by a mutation inactivating the repressor, a transcription-blocking protein, or the specific DNA sequence (operator) to which the repressor binds, but occasionally the specificity of induction is altered. Such regulatory mutations are clearly adaptive, since they allow growth and reproduction on a new source of carbon or energy.

The demonstration of the adaptive value of regulatory change is more difficult in eukaryotes because changes are not usually the result of deliberate selection. However, the precision of expression of genes that are specific to developmental stage or particular tissues, or in response to environmental signals, is frequently important in adaptation. For example, a gene in ruminants and primates is expressed in the foetus, where it makes a haemoglobin with a higher oxygen affinity than the haemoglobin of the mother, thus ensuring that oxygen is transferred across the placenta (Perutz, 1983). Adult primate haemoglobin, however, requires a lower oxygen affinity to permit a larger fraction of the oxygen to be discharged in the tissues, and it is coded by a different adult-specific gene. In the evolution of this haemoglobin system, it is probable that an original gene, now expressed in adult tissue, was duplicated and translocated. One gene copy accumulated nucleotide substitutions, in particular one leading to the replacement of serine by histidine, thus giving increased oxygen affinity, followed by the addition of signal sequences for foetal or adult expression. The evolution, following duplication, of divergence of gene expression within and among tissues is of common occurrence (Ferris and Whitt, 1979).

2.2.8 *Location of genetic changes affecting regulation*
Regulatory genetic change that affects gene transcription, as indicated by shifts in tissue- or stage-specificity, may be located either *cis*- or *trans*- to the coding sequence that specifies the protein. If the alteration is in the *cis*-position, it is on the same chromosome, usually in a protein-binding site; if it is in the *trans*-position, it may be on a different chromosome, usually in the diffusible protein that binds to a DNA site. Whether mutations are *cis*- or *trans*-located can often be determined in a first generation hybrid. If the genetic change affecting the pattern of gene expression is in a *cis*-acting site, the two alleles at an enzyme-coding locus will be differently expressed in a heterozygote. A change in a *trans*-acting factor, in contrast, will equally affect the enzymes coded by both alleles and there should be no variation between tissues or developmental stages. Finally, quantitative alterations in gene expression will follow the patterns outlined in Figure 13.

By such means, Dickinson (1980) established that the marked differences in the pattern of expression of alcohol dehydrogenase and aldehyde oxidase between *Drosphila heteroneura* and *D. differens* are caused by genetic changes located *cis*- to the relevant genes. When this type of analysis was extended to the tissue- and stage-specificity of expression of five enzymes in 13 tissues of 27 species of *Drosophila*, the major differences (11 out of 17) between species were found to be definitely due to changes in *cis*-acting genetic elements (Dickinson, 1980). Three other differences of the 17 analysed were of intermediate expression in hybrids and so might be more complex. The final three differences were due to changes in *trans*-acting (diffusible) factors, since the pattern characteristic of one parent dominated in the hybrid. Alterations in *cis*-acting flanking sequences are also compatible with most of the tissue-specific changes in gene expression in fishes (Kettler et al., 1986) and with the occurrence of δ-globin polymorphism in man (Maeda et al., 1983).

	Cis-directed change (linked)				Trans-directed change (unlinked)			
	Sequence			Activity	Sequence			Activity
	Trans	Cis	Coding		Trans	Cis	Coding	
Parent A				High				High
Parent B				Low				Low
F_1				Intermediate				One parent dominant

Figure 13 The determination of the nature of transcription-regulating genetic changes. The enzyme activity of the F1 hybrid will demonstrate whether the change is located *cis-* or *trans-* to the coding gene, according to the diagram. If the alleles differ electrophoretically in the enzyme code, the absence of rarity of recombination in the F2 or backcross generation is indicative of a *cis*-location for the regulatory change. Crosses mark the positions of genetic changes.

The significance for evolution of genetic changes in *cis*-acting regulatory sites that control the expression of specific genes is that selection for slight developmental changes can be made without disrupting existing developmental patterns.

2.2.9 *Mechanisms of transcriptional control in animals*

The importance of the regions flanking genes in the programmed release of genetic information during ontogeny becomes apparent when we consider how the transcription of most genes is regulated.

Since the direction of transcription is always 5' to 3', each gene must have sequences (promoters) preceding it, at which RNA polymerase can attach and initiate transcription. This system for independent gene expression provides the simplest level of control and forms the basis of specificity in gene action. In eukaryotes, three functional components of a promoter have been distinguished: the initiator, the selector and the modulator (Grosschedl and Birnstiel, 1980; Figure 14).

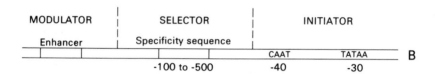

Figure 14 Diagrams of promoter structure: A, in prokaryotes, B, in eukaryotes. The bacterial promoter usually consists of two elements situated within about 50 nucleotides 5' to the transcriptional start, to both of which RNA polymerases with different sigma subunits and viral encoded polymerases probably bind. The sequences differ according to whether the polymerase is for genes transcribed in response to heat shock, nitrogen limitation, sporulation etc. Many eukaryotic promoters contain initiation sequences like prokaryotes, and also selector sequences for specific expression and modulator or enhancer sequences. There is no clear distinction between tissue- and stage-specific sequences or 'upstream elements' and enhancers.

The initiator determines the start point of transcription and comprises the cap site where a methylated guanine nucleotide is added and two short DNA sequences. One of them, the TATAA sequence, is situated 20 to 30 nucleotides 5' to the initiation site and directs RNA polymerase to the correct transcriptional initiating start site. The other is the CAAT sequence which, when present, is usually about 40 nucleotides 5' to the start of transcription and acts in conjunction with other sequences in initiating transcription. It is uncertain whether these elements have a role in the regulation of transcription. Some genes, called housekeeping genes, have no apparent TATA sequence, but have a high content of the nucleotides guanine and cytosine in their promoters (Dynan, 1986). Such genes show little tissue specificity of expression, and code for enzymes that perform essential metabolic functions. Examples are genes for hypoxanthine phosphoribosyl transferase, dihydrofolate reductase, adenosine deaminase and phosphoglycerate kinase.

The selector component of the promoter makes the gene specific for expression in different tissues or developmental stages and for induction by stress or metal ions or hormones and for the viral induction of α- and β-interferon genes. Some genes, like the one for heat-shock protein 70 in human cells, contain two distinct regulatory domains (Wu et al., 1986). One domain contains elements responsive to heat shock and cadmium, and the other an element responsive to serum stimulation of transcription. All these cis-acting elements are believed to be binding sites for trans-acting factors (proteins, complexes of protein plus hormone or metal ion etc.), which are called differentiators or differentiation inducers. The heat-shock or stress-induced genes, for example, are ones which contain regulatory elements in their promoters to which stress-induced transcription factors bind and affect gene expression. Usually in response to a heat shock or other environment stress like ethanol, anoxia and heavy metals, normal protein synthesis is shut down and there is induction of a small set of heat-shock proteins which are thought to protect the cell against the existing heat shock as well as subsequent ones. This genetic protective system has been found in all prokaryotes and eukaryotes examined and has been highly conserved in evolution; e.g. the Drosophila heat-shock gene is still induced by heat shock when transferred to pig, mouse and monkey cells (Bienz and Pelham, 1982). The induced metabolic stresses cause the partial denaturation of intracellular proteins, and these abnormal proteins activate the heat-shock genes (Ananthan et al., 1986).

The modulator component of a eukaryotic promoter corresponds to the enhancer, and modulates the frequency of transcriptional initiation; it has little or no cell-type specificity in viruses. It may be up to 6000 nucleotides distant from the coding sequence of a gene and may be either 5' or 3' to it. Some enhancers, as defined above, may be tissue- or hormone-specific in animals while others determine the level of gene expression.

It may be equally important in regulation to have a means of switching off genes without damaging them. Such negative regulation of transcription is effected by silencers which, like enhancers, act relatively independently of orientation and position in mammals (Laimins et al., 1986) and yeast (Brand et al., 1985). The mammalian silencer sequences are members of long interspersed repetitive sequences present in more than 50 000 copies in the rat genome. They are thought to have a role in regulating the expression of adjacent genes since they also carry transcriptional enhancers (Cooper et al., 1984).

The short cis-acting sequences are usually in multiple copies near or in the gene: e.g. the progesterone-receptor binding site is reiterated throughout the ovalbumin gene locus (Mulvihill et al., 1982). Generation of enhancer activity by the multiplication of single elements has been demonstrated for the SV 40 viral enhancer (Weber et al., 1984), and such multiplication is probably of widespread occurrence. These sequences provide sites, seven to 18 nucleotides in length, appropriate for the binding of proteins such as variants of RNA polymerase and other trans-acting nonpolymerase transcriptional factors like specific enhancer-recognising proteins. In the case of hormone-activated transcription, the hormone combines with specific receptor protein within the cell, enabling the complex to bind in turn with increased affinity to specific sites in the cellular DNA (Figure 15). A single protein hormone can thus control the expression of a whole battery of genes.

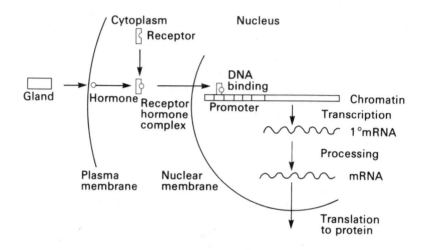

Figure 15 Diagram showing the hormonal activation of gene transcription.

RNA polymerase II, the enzyme responsible for messenger RNA synthesis in eukaryotes, differs from prokaryotic RNA polymerase in being unable to recognise promoters in *in vitro* reactions. Other proteins from cell extracts must be added to the polymerase for it to initiate transcription at all promoters, while yet others impart site-specificity to the polymerase binding. One of these *trans*-acting factors is the promoter-specific protein Sp1. This protein enables the polymerase to bind to promoters and initiate *in vitro* transcription of certain genes in the viruses SV 40 and herpes simplex, and in the monkey genome (Dynan and Tjian, 1985). All Sp1 binding regions contain one or more copies of the hexanucleotide GGGCGG, which may be present in either orientation with respect to transcription. For the transcription of the 5s RNA genes of *Xenopus* by RNA polymerase III, at least three additional protein factors are required (Miller et al., 1985). Of a similar nature to Sp1 is the heat-shock transcription factor that is involved in initiating the transcription of certain heat-shock genes (Pelham, 1982). Many more otherwise uncharacterised *trans*-acting transcriptional factors are known or suspected. The binding of these proteins to polymerases or to sites in the promoter allows the transcriptional machinery to be discriminating in the expression of genes.

There appear to be further, as yet little known, factors that contribute to the differential spatial and temporal expression of genes. This is indicated by, for example, the fact that heat-shock genes cannot be activated in the early embryo of most, if not all, animals, because the organisation of the DNA is believed to be unsuitable.

2.2.10 *Transcriptional control in other organisms*

Enhancers of transcription are not known in bacteria, but enhancers of recombination, both generalised (Dower and Stahl, 1981) and site-specific (Craig, 1985), are fairly common (about 1000 generalised recombination enhancers per chromosome in *Escherichia coli* (Smith, 1983). The cells of prokaryotes have primarily to adjust to changes in external nutrient supply, whereas eukaryotes must regulate gene expression during cellular differentiation, tissue development, organ formation and in the control of behaviour.

To illustrate the difference between prokaryotic and eukaryotic gene regulation, we can consider the control of expression of a typical metabolic enzyme, ß-glucuronidase, in the two kingdoms. In the bacterium *Escherichia coli*, ß-glucuronidase is the first enzyme of the hexuronide-hexuronate pathway. Synthesis of the enzyme is negatively controlled by two repressors which bind to an operator sequence preceding the structural genes and by catabolite repression via cyclic adenosine monophosphate. The regulatory (promoter) region of the ß-glucuronidase gene contains probable binding sites for repressor protein, catabolite activator protein, RNA polymerase and ribosomes (Blanco et al., 1985).

ß-glucuronidase in the mouse is inducible by steroid hormones, localised in intracellular position and regulated in level of synthesis (Paigen, 1979), but all forms appear to be derived from a single structural gene. The enzyme acts to hydrolyse mucopolysaccharides and steroidal glucuronides in the lysosomes of most tissues. Particularly in the liver and kidney, ß-glucuronidase is combined with a nonspecific anchor protein (egasyn) which attaches it to the endoplasmic reticulum. Unlike the simple on-off transcription switch for bacterial ß-glucuronidase, the gene in mammals is controlled quantitatively in rate of expression in response to testosterone and qualitatively by promoter sequences or genes that confer tissue specificity. Other genes elsewhere in the genome govern the secretion of ß-glucuronidase into the urine.

The sequences of promoters of plant genes are very similar to those of animal genes in that a TATA and often a CAAT sequence are essential for gene expression, and some plant promoters contain the core animal enhancer GTGG(AAA/TTT)G (Doyle et al., 1986). It has been demonstrated that certain hormones can induce the transcription of specific genes in plants: for example, the initiation of transcription of the gene for α-amylase in the aleurone of the barley grain in response to gibberellic acid (Jacobson and Beach, 1985). However, a comparison of the nucleotide sequences of the promoters of three genes induced by gibberellic acid revealed no regulatory elements common to them (Baulcombe et al., 1986).

According to Nagl (1979), the amount of condensed chromatin appears to differ among plant species, rather than in different tissues as in animals. In other respects, such as the interchangeability of plant and animal histones in forming chromatin, the structure of chromatin appears to be the same in plants and animals.

A recent review of the causes of evolution in plants by Stebbins (1986) concluded that most change was due to the alteration of proteins associated with intracellular membranes and with the cytoskeleton. The proposed focus of change in plants is in *trans*-acting factors rather than in the *cis*-acting sequences of animals. It seems to be still unclear whether changes in protein binding in sites flanking structural genes are an important source of adaptive differences in flowering plants.

2.2.11 *DNA organisation and gene expression*

Some genes can be in a potentially active state, that is to say inducible, in some tissues but not in others. For example, the growth hormone gene of the rat is induced in anterior pituitary cells by glucocorticoid hormone, but no growth hormone can be so induced in liver cells although the growth hormone gene is present (Yamamoto, 1985). In fact, it has been found that genes which are expressed constitutively in all cells of the organism and genes that are specific to particular cell types have chromatin of a different 'active' conformation from the majority of genes (McGhee and Felsenfeld, 1980). The activation of a gene does not usually result in transcription but only in conferring an ability to be transcribed in response to other signals.

Active (transcribed) or potentially active genes are preferentially sensitive to digestion by deoxyribonuclease I and contain short sequences which are hypersensitive. Deoxyribonuclease-hypersensitive sites have been observed in regions preceding genes transcribed by RNA polymerase II in fungi (Baum and Giles, 1985), plants (Eissenberg et al., 1985) and animals (Thomas et al., 1985). These sites appear to correspond to enhancers at which *trans*-active transcriptional factors bind. Factors present in erythrocyte nuclei of the fowl have been shown to interact *in vivo* with the nuclease-hypersensitive domain 5' to the adult ß-globin coding sequence to induce transcription (Emerson et al., 1985). Hypersensitivity to deoxyribonuclease is believed to reflect peculiarities of the nucleotide sequence, in particular, sequences of alternating purines and pyrimidines which are potential regions of left-handed Z DNA.

Control of transcription other than by *trans*-acting proteins and their binding sites in the flanking sequences of genes may include degree of methylation, extent of coiling, presence of histone and acidic protein complexes and perhaps the domain organisation of chromatin. These additional controls are not properly understood, and they have not yet been implicated in alterations of gene expression during evolution.

2.2.12 *Mechanisms for changing signal sequences of genes*

The short flanking sequences that make a gene specific in expression change frequently in evolution, but it is not yet apparent how this is done. Baulcombe et al. (1986) have pointed out that there are at least three ways in which gene expression at the messenger RNA level has been made tissue- or organ-specific (Figure 16). The apparently most common way, as exemplified by glutamine gene expression in *Phaseolus*, has two or more genes with the same coding sequences but with promoters responsive to different *trans*-acting inducers (Cullimore et al., 1984). A second way, as in the gene for α-amylase in the mouse (Young et al., 1981), is to have a single coding sequence but with alternative promoters that direct differential excision and splicing in the primary messenger RNA to provide different proteins in different tissues. The third way is found in the lysozyme gene of the fowl (Thiesen et al., 1986). This has a single coding sequence and a single promoter, but different regulatory elements for different cell types so that one element is for expression in macrophages and another for oestrogen control in the oviduct.

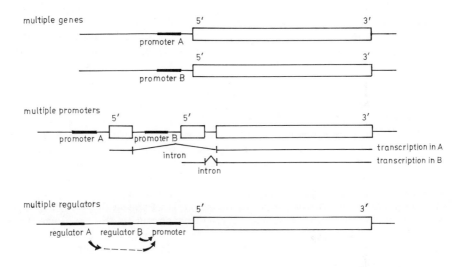

Figure 16 Strategies for tissue- or organ-specific gene expression (Baulcombe et al., 1986). In the 'multiple genes' system, the genes have similar sequences but different promoters. The 'multiple promoters' system has a single coding sequence but alternative promoters. The 'multiple regulators' system has a single coding and promoter sequence but the promoter may be controlled by different inducers.

Mutation by nucleotide substitution is unlikely to be effective in altering specificity of gene expression, at least in the usual case where specificity relies on multiple copies of short *cis*-acting nucleotide sequences. Moreover, a partial change would result in a mixing of gene expression like the formation of ß-globin in blood-forming cells and in nerve cells.

The overall recombination frequency in higher eukaryotes is about 10^{-5} per 1000 nucleotides of DNA (Paigen, 1986). Therefore, the entire gene complex of structural and *cis*-regulatory elements tends to segregate in populations as a single unit. Although the frequency of recombination varies for different chromosomes and for different regions of the one chromosome, a recombinational alteration of signalling sequences may be very rare.

A mechanism of genome rearrangement often proposed is the transposition of movable elements. Some transposons, retroposons and retroviruses have long terminal repeats containing initiation, termination and enhancer sequences. For the purpose of altering signalling sequences, the Ty movable element of yeast seems particularly suited. It is repeated about 35 times per haploid yeast genome and has a low frequency of transposition of about 10^{-8} per gene. It provides a region of homology for recombination by the normal recombination system at a frequency about 100 times the transposition frequency (Fink et al., 1986). This movable element contains an enhancer-like sequence in its long terminal repeat that can activate silent genes (Roeder et al., 1985). It transposes in the order DNA RNA→DNA, unlike bacterial elements which probably transpose DNA→DNA and retroviruses which transpose RNA DNA. Like many retroviruses, Ty elements are not infectious, but remain intracellular. Once transposed to a new position, a Ty element may undergo recombination between the long terminal repeats to excise the transposon, leaving a single long terminal repeat (Winston et al., 1984). The frequency of this event is about 10^{-5} per element or 1000-fold higher than the transpositional frequency.

The copia-like transposable elements of *Drosophila* are present in about 30 different families and account for nearly half of the moderately repeated DNA (Young, 1979). They are structurally analogous to the Ty elements of yeast and to the integrated proviruses of the vertebrates. Unlike the long terminal repeats of Ty elements in yeast, which are present much more frequently in the genome than complete elements (Cameron et al., 1979), those of copia element are not found separate from the main body of the element (Levis et al., 1980). Nor are the other *Drosophila* elements good candidates for evolutionary agents. The long terminal repeats of gypsy elements, for example, contain multiple nonsense codons in every possible reading frame and have no promoter or functional donor or acceptor splicing sites at their ends (Freund and Meselson, 1984.

The transposable elements of maize, such as En/Spm, Ac/Ds and Mu 1, as well as Tam of *Antirrhinum majus* and Tgm of *Glycine max* all lack terminal transcription signals. However, the integration of *Mul* in the 5' flanking region of the gene for alcohol dehydrogenase in maize results in a great reduction in the enzyme expressed in the seed but not in the pollen (Chen et al., 1987). The insertion creates an additional TATA sequence but it is not known if this is connected with the alteration in organ-specific expression.

Retrotransposons are not known in vertebrate cells, but retroviruses are common. In mice there is a family of about 500 copies per genome of repeated sequences resembling solitary retroviral long terminal repeats that can apparently recombine with exogenous retroviruses (Schmit et al., 1984). Also in mice are retrovirus-like elements known as intracisternal A particles coded for by genes that are reiterated 1000 or more times per haploid genome. These elements carry long terminal repeats of 350 to 400 nucleotides that are closely homologous in sequence to elements found in the promoter of an activated cellular oncogene and in an intervening sequence of a defective K light-chain gene (Kuff et al., 1983). However, the intracisternal A particles are unlikely to be agents of mammalian evolution because, although their genes occur in many copies in Syrian hamsters, they are apparently absent in Chinese hamsters (Temin, 1985).

All inbred strains of mice have a gene that specifies the synthesis of rennin in kidney cells and at a relatively low level in the submaxillary gland. Certain strains also synthesise an isozyme of rennin with up to 100 times the usual activity in the submaxillary gland from a closely linked duplicate gene. It appears that a gene normally expressed in the kidney and weakly so in the submaxillary gland has been duplicated to provide a copy that is expressed only in the gland. A comparison of the sequences of the promoters of these duplicate genes indicates that, as well as a few other differences, a mouse type 2 Alu-like element has been inserted in one of the promoters about 170 nucleotides prior to transcriptional initiation. This insertion may be responsible for the tissue specificity of the duplicate gene (Field et al., 1984), but its action is not clear.

2.2.13 *Conclusion*

It is evident that an important evolutionary mechanism is one that changes a gene so that its expression at particular times or places is altered. Such a mechanism would probably need to be confined to single genes because changing the specificity of expression of batteries of genes could grossly derange normal development. For this reason, changes in the genetic system should not affect such manifoldly acting components of the gene switching system as the environmental signal receptor, the hormonal messenger or the membrane or intracellular hormone receptor. As we have already seen, specificity in transcription is provided by DNA sequences around, and particularly preceding, the protein-coding part of a gene. In addition to the RNA polymerase and ribosome binding regions, other short discrete sequences provide the *cis*-located information for enhanced or reduced transcription, tissue- or stage-specificity, heat and other stress responses and possibly for other factors of the internal and external environment. As with domains in proteins, these small DNA sequences are arranged on a modular basis in that each is separated by a nucleotide sequence of apparently only spacer function. Therefore, the regulatory elements of eukaryotic genes need not coevolve, and if separately transposed they need not be in any strictly determined position or orientation. This positional flexibility of the specificity components of many genes suggests that an alteration of signal sequences could promote radical evolutionary change without many adverse side effects.

It is not yet possible, however, to say how the signals for expression are altered during evolution. Examples of transposable elements activating or altering gene expression have been noted in bacteria, yeast, maize and mouse, but these may have been only incidental events. They do, however, indicate a high sensitivity of sequences flanking genes to breakage, insertion, deletion etc. In general, the nuclease-hypersensitive regions preceding transcriptional initiation are sites that are particularly liable to DNA cleavage agents, including deoxyribonucleases I and II, micrococcal nuclease, restriction endonucleases and chemical cleavage compounds. In addition, they are preferential sites for recombination and transposition; of the 70 000 nucleotides encompassing the human ß-globin gene cluster, 75% of all meiotic recombinants occur in a 9100 nucleotide sequence 5' to the gene (Chakravarti et al., 1984). Also, Maeda et al. (1983) have found that there are sequences within the 3100 nucleotide region 5' to the ß-globin gene that predispose it to recombination. With regard to transposition, the Ty element of yeast inserts preferentially into the 5' region of the lysine 2 gene (Eibel and Philippsen, 1984) and Simchen et al. (1984) found that nine out of ten transpositions were to this region. This selectivity of insertion is believed to rest on a higher affinity of nucleases for opened regions in chromosome structure like those of regulatory sequences.

Another factor that must be taken into evolutionary account is the gross structure of the chromatin. When, for example, murine mammary tumour virus is integrated at different places in rat hepatoma chromosomes, it appears to adopt the structure of the flanking cellular chromatin. Thus a provirus located in nuclease-sensitive chromatin was transcriptionally induced by glucocorticoid hormone, while another in nuclease-resistant chromatin was not expressed (Feinstein et al., 1982).

2.3 *Formation of Higher Categories*

The rate at which evolution progresses, and the direction that it takes, can only be reliably assessed from fossil remains, so it is mainly from the shells of invertebrates and the bones of mammals in the Tertiary that the course of evolution has been reconstructed. The conclusions really apply only to the animal taxa providing the fossil data, but they have been extrapolated to include prokaryotes, fungi and plants.

The taxonomic groups here considered, classes and orders, generally represent distinct adaptive types by virtue of the fact that they possess one or more novel features of structure or function. The new characters they possess are usually different in kind and degree from those that are typical of new species, genera and perhaps families. However, following the origin of a major group, it is commonly diversified into lower categories adapted to specific environmental subdivisions. As this occurs, genetic changes of the same type as those that characterise specific and generic differences accumulate.

There are two views as to the origin of the innovations determining class and order designation. One is that the character concerned has been built up slowly by numerous small mutations either affecting the character directly or having a pleiotropic influence on it. The other view proposes that the genetic changes leading to the new ordinal character are different from those causing generic and specific diversification. Of course, both views concede that the new structure is arrived at by modification or elaboration of a pre-existing one. And most would agree that, if there are two types of mutation, both may occur in the generation of diversity, whatever the taxonomic level.

Several objections have been raised against the belief that a buildup of mutations (point mutations and chromosomal rearrangements) are responsible for the origin of such novelties as characterise a class or order. One is that their appearance is said to be much too rapid to be accounted for by such mutational events even with rigorous selection. Another objection to gradualistic evolution is that where numerical changes are involved, such as number of segments, vertebrae or heart chambers, relatively large morphological steps require to be invoked. Finally, the novel feature of a higher taxon may need to be evolved abruptly because in some cases it is not possible to visualise a selective or adaptive role for the partial structure or function.

Objections to the idea that the mutations initiating higher taxonomic levels are different in kind from those involved in specific differences include the claim that it is contrary to the view that genetic variability in populations provides the material of all evolution. Indeed Simpson (1953) has stated that there is no reason to conclude that the mutations leading to the distinctive characters of classes and orders are any different in degree from those separating species and genera. Furthermore, it has been considered that, apart from nucleotide substitution, deletion etc. in point mutations, no mechanism exists at the genetic or molecular level which could lead to viable mutations of large effect. With respect to this major objection as to a lack of mechanisms for large mutations, it is now realised that, in addition to point mutations of small effects, there are others of a 'recombinational' type which can at least produce new genes. The details of these genic and subgenic rearrangements are considered in Part 3. In the present Section, we will confine our attention to the three kingdoms for which reasonable evolutionary information is available: prokaryotes, plants, and animals.

2.3.1 *Prokaryotes*

Despite the conclusion that evolution has for several eras acted primarily through developmental change, this cannot hold for prokaryotes with their very limited capacities for development. The principal characters defining the four main prokaryotic subgroups (cyanobacteria, myxobacteria, spirochaetes and eubacteria) are indeed morphological and they have been used in outlining evolutionary trends (Stanier and Van Niel, 1941). But as regards cell shape (cocoid, rod-like etc.), it is uncertain whether the wall merely follows the membrane contours or its assembly determines the shape. It is possible that the final shape of many bacterial cells is solely a consequence of such factors as surface tension, osmotic pressure, mechanical restraint and compositional discontinuities in the cell envelope (Henning, 1975).

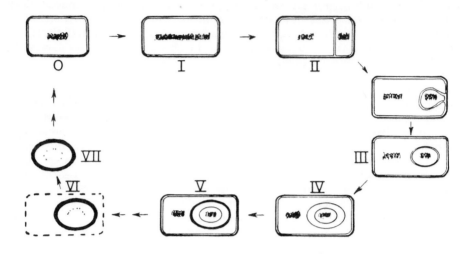

Figure 17 Stages of spore formation in Bacillus subtilis (Piggot, 1985).

Stage 0	The vegetative cell.
Stage I	Condensation of chromosome into an 'axial filament'.
Stage II	Formation of membrane septum.
Stage III	Migration of septum to engulf and pinch off the forespore.
Stage IV	Deposition of the cortex around the spore protoplast.
Stage V	Formation of spore coat.
Stage VI	Maturation of spore coat.
Stage VII	Release of spore by lysis of mother cell.

Some prokaryotes exhibit ordered processes that are usually regarded as developmental because they are defined sequences of events which result in an alteration in morphology. One of the best studied is the formation of spores in *Bacillus subtilis* (Figure 17). Like programmed developmental gene expression in other microbial systems, the sequential activation of sporulation genes is triggered by nutrient depletion. An early step in spore formation is the invagination of the plasma membrane to form a septum dividing the cell into a mother cell and a forespore compartment. The forespore develops a cortex and a thick protein coat and is released as a dormant spore when the mother cell lyses (Fitz-James and Young, 1969). These changes to the vegetative cell are the result of a differential transcription of new genes and a cessation of transcription of many vegetative genes. The transcriptional changes are brought about by a replacement of one of the subunits (the sigma factor) in the RNA polymerase active in vegetative growth by a different one when sporulation begins. Subsequently, this sigma subunit is replaced by other stage-specific ones during sporulation to transcribe the relevant genes in ordered progression (Losick and Pero, 1981).

The existence of multiple polymerase forms that are utilised for the transcription of different gene sets commonly occurs as a feature of the transcriptional machinery of bacteria. When the cells of *Escherichia coli* are shifted during growth from a low to a high temperature, the synthesis of certain proteins increases transiently (Neidhardt et al., 1984). Genes for such proteins (heat-shock genes) contain specific DNA sequences in their promoters which, at least in one instance, are homologous to that of a heat-shock gene in *Drosophila* (Bardwell and Craig, 1984). The switch to enhanced heat-shock gene transcription is brought about by an RNA polymerase in which a heat-shock sigma factor replaces the former factor. Similarly, the transcription of genes controlled by nitrogen availability in enteric and nitrogen-fixing bacteria is determined by an RNA polymerase containing a sigma subunit specific for promoters related to nitrogen-regulated genes (Magasanik, 1982).

The sequencing of bacteriophage and bacterial genes from *Bacillus subtilis* and *Escherichia coli* that encode sigma factors show that they are all members of a family of partially homologous genes. All sigma proteins sequenced are homologous to each other in a highly conserved region thought to be the site of binding of these factors to the core polymerase (Gribskov and Burgess, 1986).

Evidently, the strategy for differential gene expression used for metabolic adjustment to environmental change has been applied to a simple developmental process involving spore formation in response to similar environmental influences. In all these instances, differential gene transcription occurs within the single cell, except that gene expression is segregated to some extent between the forespore and mother-cell compartments. Development in multicellular organisms, on the other hand, must distinguish between cells that are to undergo differential transcription and cells that do not.

Among the prokaryotes are examples of cellular aggregation, multicellularity and cellular interaction. For example, the myxobacteria live as multicellular organisms in that they move and feed in organised masses and they undergo a multicellular developmental process in forming fruiting bodies. The species lives in soil as aggregates of cells which obtain their food from insoluble macromolecules like cellulose, microbial cell walls and proteins that are degraded by secreted hydrolytic enzymes. The efficiency of feeding has been shown to rise with increasing cell density (Rosenberg et al., 1977), thus accounting for the evolution of the cells to live in masses. When nutrients become depleted, a period of multicellular morphogenesis (aggregation into species-specific shapes) is followed by a period of cellular morphogenesis (change of rod-shaped vegetative cells into round, resistant myxospores). The signal for aggregation has been shown to be adenosine that is secreted into the

77

medium by starved cells (Shimkets and Dworkin, 1981). During aggregation, further cell interactions, including mass movement, are mediated by cellular contact. During the development of fruiting bodies, differential transcription is induced, as shown by a new population of messenger RNA

TABLE 7

Patterns of feedback inhibition of the first enzyme in bacterial aromatic amino-acid biosynthesis; i.e. 3-deoxy-D-arabino-heptulosonate-7-phosphate synthetase (Data of Jensen et al., 1967, as arranged by De Ley and Kersters, 1975). With few exceptions, a pattern of enzymatic control distinguishes bacteria at the generic level.

Type of Feedback Inhibition	Occurrence
(a) Sequential Feedback inhibition	*Bacillus, Sporasarcina, Staphylococcus, Gaffkya, Flavobacterium, Achromobacter parvulus, Alcaligenes visolactis*
(b) Isoenzymatic feedback inhibition dominant phenylalanine-sensitive isonzyme	*Escherichia, Erwinia carotovora, Aerobacter, Serratia, Erwinia amylovora, Aeromonas*
mainly phenylalanine- and tyrosine-isoenzymes	*Salmonella, Shigella, Saccharomyces cereviseae Neurospora crassa*
(c) Concerted or multivalent feedback inhibition	nonsulphur purple bacteria
(d) Cumulative feedback inhibition	*Hydrogenomonas*
(e) Synthetase inhibited by tyrosine	*Pseudomonas, Neisseria, Mycobacterium, Nocardia*
(f) Synthetase inhibited by phenylalanine	*Alcaligenes faecalis, Achromobacter viscosus, Veillonella*
(g) Synthetase inhibited by tryptophan	*Myxococcus, Streptomyces*

molecules with half-lives of 20 to 30 minutes as compared with three minutes for the messengers of vegetative cells (Nelson and Zusman, 1983). An almost identical set of developmental stages occurs in the eukaryotic slime mould, *Dictyostelium* (Loomis, 1982).

The particular interest of *Myxococcus* is that it provides evidence of the prokaryotic origins of many of the developmental processes of multicellular eukaryotes, such as cellular differentiation, aggregation and movement. The interaction between cells that is a prerequisite for the multicellular conditon may have evolved from the chemosensory systems that enable single cells to detect food and avoid toxins (Carlile, 1975). Myxobacteria may represent one of the earliest organisms to attain multicellularity which, according to their 16S ribosomal RNA sequences, originated about 2000 million years ago when the atmosphere is thought to have become aerobic (Kaiser, 1986).

Evolution, at least in the later evolved prokaryotes, has been concentrated primarily on the control of intracellular metabolism. While metabolic pathways are usually the same in different bacteria, the pattern of regulation may show considerable variation, as might be expected in later evolved functions. Jensen et al. (1967) examined the control of aromatic amino acid biosynthesis in 91 species of bacteria belonging to 32 genera. At least seven different control mechanisms based on feedback inhibition occur (Table 7). This enzyme control system varies as do others that have been compared in bacteria, like that for citrate synthase (Weitzman and Dunmore, 1969); i.e. the same control patterns are usually found for all species of a genus although the difference between families may be considerable.

2.3.2. *Plants*

The plant and animal kingdoms have evolved separately for more than 1300 million years when the first known alga is found in the fossil record. Thereafter, a difference in nutrition, photosynthetic in plants and predatory in animals, is correlated with a sessile life-style in the former and a mobile one in the latter. On coming to land, the plants adapted to a fluctuating environment by regulating their development, while animals adapted mainly by modifying their behaviour. Consequently, plants have evolved a flexible type of ontogeny so that alterations can be made during growth, while animals evolved a strictly programmed sequence of successive transformations with a centralised control system. In place of an animal's sequential programme, plants use external environmental signals to initiate the onset of developmental processes. This ensures that stages of the life-cycle like germination and flowering take place when environmental conditions are most favourable, and that metabolism and growth are reduced when environmental stresses occur. The advantage of reversibility in plant development is achieved by the evolution of meristems (growing points) distributed throughout the plant and the avoidance of assembly processes that increase in complexity as development proceeds. The eventual result has been that plants have evolved a developmental programme relatively resistant to disruption by mutation, whereas most mutations affecting animal development are highly pleiotropic and frequently lethal.

Another consquence of the developmental difference between plants and animals is the greater tolerance of plants to chromosomal changes, e.g. to polyploidy, supernumerary chromosomes and aneuploidy. In contrast, mammals can tolerate only certain chromosomal translocations, trisomics of the sex chromosomes and of one of the autosomes (chromosome 21) and aneuploidy of the sex chromosomes.

Developmental patterns

Although most plants, except some algae, are multicellular, their level of development is as far inferior to that of the more advanced animal as bacterial development is to that of the plant. Differentiation, as expressed in the number of cell types that are genetically programmed, produces only about 19 varieties of major cell, unlike the mammalian body which has more than 200 (Table 8). Consequently, the number of plant tissues, usually composed of more than one type of cell, is also limited. Moreover it is rare to find in plants such markedly different internal compositions as are typical of the cells of liver, muscle, nerve, pancreas etc. in animals. The cellular aspect of differentiation was evolved early, and there is no evidence that new cell types are now produced.

TABLE 8

Estimates of the number of differentiated cell types in various organisms.

Organisms	No of Cell Types*	References
Gonium (alga)	1	Kochert (1973)
Volvox (alga)	2	Kochert (1973)
Dictyostelium (slime mould)	3	J. Williams (unpublished)
Mushroom	4	Bonner (1965)
Sycon (sponge)	6	Greenberg (1959)
Hydra (coelenterate)	9–10	Markert & Ursprung (1971)
Flowering plant	19	Bloch (1965)
Earthworm	66	Stebbins (1982)
Insect	100–150	Stebbins (1982)
Whale	200	MacLean (1976)

*Excluding spores and gametes

In plant morphogenesis, altered organisation can occur by changes in meristem activity with respect to rates of cell division and growth hormone production, local alterations in planes of cell division and shifts in the position and extent of hormone induction in vascular and other tissues. Morphogenic cell movement only occurs in the very early embryo. When rigid cell walls developed around the boundary membrane, the possibilities of cell adhesion were removed and morphogenesis by cell migration, as occurs with the plastic cells of animals, became impossible. In place of cell movement, tissue differentiation can be brought about in plants at localised places under the influence of hormones and other morphogenic substances. Embryonic cell lineages appear to have highly specific developmental fates, but there is no evidence of a determinate pattern of cell division (Poethig et al., 1986).

It is notable that in the evolution of plants there have been many rather unrelated lines of advance during which their characteristic features of organisation have arisen recurrently and independently. Even allowing for the rarity of early plant fossils, there is no evidence that evolution in the plant kingdom was solely a single trend towards increasing complexity. The coincidence of certain aspects of organisation in plant groups which are little related appears to be due to their convergent or parallel evolution, which are very common phenomena in plants. So the same types of vegetative habit such as succulent, climbing, epiphytic, parasitic, aquatic, herbaceous and arborescent repeatedly appear in unrelated groups (Wardlaw, 1965). Closely analogous features presumably arise repeatedly because the simplicity and limited potential range of plant structure relative to that of animals restrict the extent to which evolutionary change can take place.

It is usual, in plants, that an altered environmental factor signals the onset of a developmental process; in animals it is the signal for a change in behaviour. Environmental triggers regulate various adaptive responses in plants, such as the timing of germination and flowering when environments are favourable and the modification of the phenotype by leaf loss etc. under stresses like cold and drought. Certain phenotypic characters can undergo as large a change from an alteration of the environment as from an alteration of the genotype. Examples are the completely different leaf shapes that may result from initiation above or below the water surface and the distinctive juvenile and adult foliage of some trees. Such alternative phenotypes may be produced by an environmental, hormonal or genetic signal. Apart from mineral dificiencies and effects on enzymatic catalysis, nearly all environmental influences on plants are mediated through hormones.

There is, however, much genetic variation in the ability of hormones to mimic the environmental signal. Gibberellic acid, for example can replace the cold requirement for flowering in *Oenothera lamarkiana* but not in *O. biennis*, and can promote short-day flowering in *Lactuca dentata* and *L. sativa* but not in *L. scariola*, all long-day plants (Zeevaart, 1978). As with other organisms, it is the control processes of development that show the most natural variation, and even when these processes are regulated by similar hormones the exact degree of timing of response depends on the genotype of the species. Thus it is evident that adaptive evolution can occur at specific developmental stages by alteration of the mechanisms by which environmental signals influence gene expression.

A major environmental factor in development is light duration and intensity. Light signals not only change the activity of the shoot meristem from the vegetative to the reproductive phase, but also induce alterations in the internode distance between leaves, influence leaf shape and alignment, initiate pigment synthesis and affect germination. At the cellular level, Tobin and Silverthorne (1985) report that more than 40 enzymes increase in activity when etiolated plants are exposed to light. One of the light receptors is phytochrome, a pigment sensitive to red and far-red light, but others exist that absorb blue and ultraviolet light. Many of the phenotypic responses to light appear to occur by differential gene expression, but others such as stomatal movement and leaf closure perhaps do not, because the speed of response seems to be too great. Responses to light like the synthesis of certain proteins active in photosynthesis occur because some genes have specific sequences in their promoters for light-regulated or phytochrome-mediated transcription (Fluhr et al., 1986). However, in general, the precision of gene expression seems to be less specific in plants than in animals. In tests of isolated genes by transformation into another species, plant gene expression is quantitatively variable but there are no reports of expression in inappropriate tissues (Nagy et al., 1985), unlike the results of comparable transformations in mice (Palmiter and Brinster, 1985).

Developmental evolution
A range of single-gene morphological mutants in plants has been examined by Hilu (1983). Many of these simply-inherited mutations affect taxonomic characters that are used to distinguish plants at the specific, generic, familial and even ordinal levels. Such are: inferior to superior ovary, few to many stamens, actinomorphic to zygomorphic perianth, unisexual to bisexual flowers, determinate to indeterminate growth, compound to simple leaves etc. It is not implied that these particular characters are determined by only one or two genes, but that the change in character can be caused by few mutations. It is not known whether the mutations responsible are point mutations, chromosomal rearrangements or other genetic alterations. A genetic and developmental system apparently exists whereby, through simple genetic change, the characteristics of a new plant taxon can arise abruptly. Probably the uncomplicated ontogeny and developmental simplicity of plant organisation allow such major morphological changes to be accepted because they do not seriously affect viability. Their occurrence suggests that it is not necessary to invoke the gradual accumulation of mutations of small effect as an explanation of much of plant morphological evolution, although small mutations of modifying and regulating genes are probably important in the developmental and environmental adjustment of the new phenotype. The nature of these mutations may account for the observation of Stebbins

(1974) that all the morphological differences that separate orders and families of flowering plants occasionally exist as differences at the specific or infraspecific level. Furthermore, it is probable that the major trends in floral evolution (polypetaly to gamopetaly, actinomorphy to zygomorphy, raceme to solitary inflorescence etc.) originated and developed independently in different families. The determination of physiological adaptive features in ecotypes or species of plants, in contrast, is often found to be multigenic (Clausen et al., 1948).

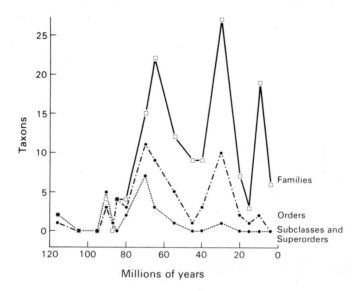

Figure 18 The times of first appearance, according to the fossil pollen record, of subclasses and superorders, orders and families of dicotyledonous angiosperms (data from Muller, 1981)

The nature of these morphological mutants is also compatible with the course of plant evolution. The diagram in Figure 18 of superorders, orders and families of dicotyledonous plants shows that the peaks and troughs of origination of the different taxa are nearly coincidental on the geological time scale. It shows a deviation from the usual pattern of animal evolutionary progression where the peaks of lower taxa usually appear following the origination of a higher taxon. The distinctive plant pattern is not due to the class being recently evolved without time to undergo much diversification, because data for the class Mammalia over much the same period show the typical animal pattern of evolution; that is, the generic peaks of origination occur 25 to 50 million years after the ordinal peaks, with the peak for families being intermediate between the two (Simpson, 1953). The inference that can be drawn is that there is no evidence in the origination of dicotyledonous plant taxa for more than one type of evolutionary mechanism, as will be suggested below for the animals.

The species of flowering plants far exceed in number those of mammals, although both classes have evolved for roughly the same time (about 100 million years). There are at least 200 000 species of dicotyledonous plants (Fuller and Tippo, 1954),as against about 4500 species of mammals (Rothschild, 1965). The constraints imposed by the vastly greater complexity of morphogenesis in mammals may lessen the probability of chance differentiation, and consequently fewer species have been formed.

2.3.3 *Animals*
There is one phylum of invertebrates, Echinodermata, which has a very reliable fossil history, perhaps the best presently available. This is because a high proportion of echinoderms occur in shallow seas where they are easily recognisable from their characteristic exoskeletons; most of their basic functions are represented in their skeletal features, and their taxonomy and morphology are very well known. In addition, Campbell and Marshall (1987) concluded that their fossil record is so complete that the prospect of finding major new morphological designs is now remote. For these reasons, the history of the echinoderms presents the most reliable data there are for the course of evolution.

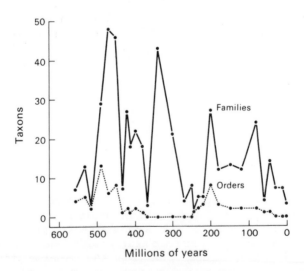

Figure 19 The times of first appearance of orders and families, both extinct and extant, of echinoderms according to the fossil record (data from the Cambrian to the end of the Permian from Campbell and Marshall, 1987: data from the Triassic to the Miocene from Sepkoski, 1982).

The graphs of the time of taxonomic origins in the echinoderms (Figure 19) give a pattern quite unlike those of flowering plants (Figure 18). Whereas the peaks and troughs of origination of plant taxa coincide, those of the echinoderms do not. Rather, the orders appear first and the families later. The diagrams suggest that the higher categories arise first and subsequently diversify into lower ones. Although the fossil data are not as complete, the arthropods also seem to have evolved in a similar pattern, as have the mammals. These patterns again raise the question that has been asked before, particularly by Goldschmidt (1960) and Schindewolf (1950), as to whether evolution proceeds in the direction species, genus, family, order, or in the reverse direction.

For the echinoderms, Campbell and Marshall (1987) point out that some of the classes and orders were established very quickly, in less than perhaps five million years, which is somewhat longer than most species durations but much less than those for genera (Table 4, page 36). They conclude that new classes and orders arise abruptly, not following an increase in diversification, and that they do not converge towards their time of origin. Nor are their origins associated with any sort of radiation, and they are therefore not due to the crossing of a threshold involving an innovation that enabled new environmental adaptations. These findings contradict the current model of higher category evolution which says that *trans*-specific evolution depends on the same genetic and selective factors as those that act in species formation, and that higher categories of organisms usually evolve when a relatively small number of individuals enter a new adaptive zone. Then it is believed that selection pressure is very intense, adaptive radiation is rapid and there is extinction of the immediate precursors of the new adaptive type.

The phenomenon of the early origins of major biological groups, suddenly and with distinct morphology, has also been remarked by Willis (1940), Nicol et al. (1959) and Valentine (1973). For example, 80% of flowering plant orders and 90% of living mammalian orders originated within a 20 million-year period in the Palaeocene. Valentine (1973) has attributed rapid large scale change to radiation occurring in empty adaptive zones, but Campbell and Marshall (1987) find no evidence for the open space theory or other environmental factors in the higher order evolution of the echinoderms.

It remains to consider what the mechanisms for large viable changes in animal evolution may be, because certain palaeontological data on the rapid establishment of novel structure and function seem to require them. Campbell and Marshall (1987), having carefully considered the very detailed fossil evidence for echinoderm evolution, concluded that 'there seems to be no alternative but to seek some unusual feature of the primitive genome that would allow it to change in such a way that large co-ordinated viable morphological changes could take place over short periods of geological time.'

Early embryogeny
The early stages of animal embryogenesis, including cleavage and blastula formation, are largely directed in many species by morphogenic

determinants transmitted through the cytoplasm from the maternal parent (Davidson, 1976). Such a substance is the protein from maternal sources that appears to initiate gene expression in the amphibian zygote (Brothers, 1976). However, if the gene coding for this protein in *Amblystoma* (order Urodela) is inactivated, the protein can be fully provided by extracts from the germinal vesicles of other amphibia like *Xenopus* or *Rana*, both of the order Anura (Briggs, 1972). This indicates that the transcription initiation process of gastrulation in 'mosaic' eggs has been evolutionarily conserved at the level of class.

At least in the vertebrates, genetic changes at the egg stage of the life cycle are very drastic so this is unlikely to be a point of development where evolutionary changes occur. If the eggs of agnathans (jawless vertebrates) and amphibia, organisms that have been separated by many millions of years, are compared, no structural differences can be detected which are related to the evolutionary transformations that have occurred (Devilliers, 1965).

In general, only maternal characters are expressed up to the gastrula or post-gastrula stage. But even at the level of the gastrula, the patterns of morphological areas are closely homologous (Pasteels, 1940). Some differences can be detected in the morphogenic movements of the cells in respect to extent, intensity and chronology. They affect processes that lead to an extension of the area of the ectoderm, convergence and stretching of notochord and mesoderm, and invagination of the ectoderm, but are only minor in expression. For example, the gastrulation movements are the same in reptiles and amphibians, but the time relations are a little different. Early embryonic development in the lower vertebrates apparently changed very little until about the time when the gastrula became rearranged. This was markedly modified in going from the anamniote type of embryo in fish and amphibia to the amniote type with extra-embryonic membranes in the reptiles, birds and mammals.

Cell lineages
After the formation of the gastrula, at the stage of differentiation of cell types, the genome of the embryo takes control of development. At each cell division, sister cells become different from each other and from their mother cell, due to the expression of different cell-type specific genes to constitute the cell lineage. Since identical lineages can be generated by different single precursor cells, there must be a series of specific programmes of gene expression that establish the lineages. Comparisons between the developmental programmes of the nematode genera *Caenorhabditis* and *Panagrellus* have identified four types of cell lineage transformations that have occurred during evolution (Sternberg and Horvitz, 1982). One is a change in the determined state of a cell, as when a particular cell becomes a distal tip cell in one genus but undergoes programmed cell death in another. Secondly, there may be a change in the number of rounds of division that a cell undergoes; e.g. differences in division number in the formation of the hypodermis. Thirdly, the polarity of a precursor cell may be altered so that the descendants of one of a pair

of cells change from the anterior to the posterior position and the reverse occurs with the other member of the pair. Finally, there may be a coordinate switch in the fates of two sister cells; that is, an altered segregation where there is a shift in cell fate from the descendants of one cell to those of its sister cell.

At present the indications are that cell lineages in the nematode are composed of separately programmed modular sublineages. Modifications of the cell lineage programme may involve alterations within sublineages by one of the transformations already mentioned or by the substitution of one sublineage by another. New cell lineages are thought to occur by a process of cell duplication followed by the modification of one of the duplicated cells or its progency (Sternberg and Horvitz, 1984).

At the molecular level, a subprogramme can be thought of as comprising the expression of a specific set of genes; about 22 in the formation of a nematode vulval cell (Ferguson and Horvitz, 1985). The signal for the switching-on of a gene set is usually generated within the cell and not by the influence of adjacent cells or by other external conditions. Other pairs of cells may constitute an equivalence group in which each can adopt either of two alternative fates depending upon position; i.e. their fates are not autonomous but are specified by cell-cell interactions. The nematode developmental system with its small fixed number of cells and its rigid cell lineage does not seem to apply to insects and mammals where lineage is not as important in specifying the fate of a cell as is positional information.

Cellular movement

In the early gastrula, tissue-specific cell lineages are produced in vertebrates by the first divisions, while the next divisions produce cells that undergo morphogenic movements to positions where they initiate tissues and organs. Although there has been an increase in the number of types of differentiated cells during evolution (Table 8, page 80), comparisons within large groups of organisms like the vertebrates indicate that cell types are very constant and that it is their arrangements in tissues that differ. The evolution of new tissues is also of rare occurrence and occurs by the modification of existing tissues. For example, the photoreceptor cells in human eyes and the sensory cells of the inner ear have been derived from modified ciliated or flagellated cells (Eakin, 1968).

One type of genetic control of tissue formation is probably exerted through changes in cell adhesiveness, cell movement and the ability of cells to migrate through tissues (Bonner, 1974). The known mutational examples have rather gross effects, such as the production of numerous digit primordia in the fowl or the impaired development of the brain cortex in mice. However, there will undoubtedly also be small genetic changes affecting these processes which will be expressed as minor shifts in morphology or pattern. The movement of melanophores, which are responsible for pigmentation, has been shown by Twitty (1949) to be genetically different in two species of salamander. The melanophores of *Triturus rivularis* migrate from the neural crest and become evenly spaced because of their mutual repulsion. However, those of *Triturus torosus* do not move away from the neural region but remain there to give a black streak down each side of the animal. When the melanophores are grown in culture, those from *T. rivularis* spread out evenly, while those from *T. torosus* move only short distances and tend to aggregate.

Induction

The major cause of change in tissue type and placement appears to be shifts in the induction-reaction processes of tissues. Mutations can result in a loss of the ability of tissue to respond to induction as in the *brachy* mutant of the mouse (Herskowitz, 1967) or in the reduction or loss of the inducing function of a tissue; e.g. the weakened ability of the retina to induce lens in nearly eyeless cavefish (Sadoglu, 1967). Nevertheless, in the course of evolution, the patterns of embryonic induction appear to have remained relatively constant. The conclusion from numerous transplantation experiments in amphibians is that, although the induction and reaction systems function coordinately, they are under separate genetic control and may be changed independently by mutations. Most frequently, it is the ectoderm that has altered in the determination of many species-specific changes in structure. The mesoderm and endoderm, in contrast, usually carry the features of the induction-reaction system that are common to species and, often, orders. For example, the scales of reptiles, the feathers of birds and the hair of mammals all develop from inductive interactions in the skin, which consists of an epidermis derived from the ectoderm and a dermis derived from the mesoderm. If grafts are made combining duck epidermis with fowl dermis or the reverse, the type, size and distribution of feathers are determined by the source of the dermis. As between the two orders represented by ducks and fowls, the epidermis has changed very little, but the dermis has altered. If grafts are made recombining reptilian epidermis with dermis from different classes like birds or mammals, abortive scales result which conform to the feather or hair pattern, but neither feathers nor hair result (Dhouailly and Sengel, 1973). Although the inductive activity of the dermis is still present in different classes, reactions have diverged at this taxonomic level to produce different epithelial structures.

Inductive interactions have become progressively more important as animal embryology has evolved, replacing the self-differentiation capacities of invertebrates with epigenetic interactions. A few inductions occur in nematodes, and in tunicates a major induction is of neural tissue by notochord, but in the vertebrates induction has assumed great importance in the organisation of the embryo. The interest of the induction-reaction system of tissue and organ differentiation on the evolution of animals is that it enables a visualisation of how organs can be increased or reduced in size, or disappear altogether, or be changed in position or in time of developmental appearance under the joint influence of mutation and selection. Morphological alterations of these types may constitute a major part of an animal's adaptive evolution.

Heterochrony

At the stage when many of the organs are developing as independent units, mutations are possible which alter their maturation times in relation to that of other organs and the whole body. Such mutations cause heterochrony (Gould, 1977), frequently manifest as neoteny, where somatic development is retarded but reproductive development is not, or progenesis where the timing relations are the reverse of neoteny. Other mutations of developmental timing affect the number of larval stages, and changes of this type are characteristic of different nematode species (Ambros and Horvitz, 1984). Changes in these and related processes readily occur because a coordinated developmental system can be shifted in relation to general development as a means of environmental adaptation. A well known example is the axolotl, which when sexually mature is in the larval form, thus remaining aquatic and avoiding the dessicating conditions of a dry land environment. Similarly, some flightless birds have the development of the sternum specifically delayed in environments lacking predatory mammals, thus saving the energy normally used in making and maintaining wings (Feduccia, 1980). Where one somatic feature is altered in developmental timing, it is thought that alterations in inductive interactions have occurred because of a failure of the inducing and target tissue to make contact at the proper time (Cahn, 1959).

Developmental switches

The mutations cited above were ones which led to failures in an ontogenetic process, so the problem still remains of accounting for the relatively major evolutionary changes that have apparently produced novel morphological structures. Raff and Kaufman (1983) propose that the production of such large evolutionary alterations may only require change in a relatively few key developmental genes of the regulatory type (switch genes) which give nonlethal phenotypes when mutated. The switch genes are thought to operate by providing binary choices during ontogenesis. An example by Stent (1985) is that 'in the developmental ancestry of a cholinergic motor neuron there would occur a commitment first to ectoderm rather than to mesoderm, then to nervous tissue rather than to skin, then to neuron rather than to glia, then to motor neuron rather than to sensory neuron, and finally to synthesis of acetylcholine rather than to γ-aminobutyric acid.'

Support for the switch gene proposal is given by the characterisation of a developmental difference distinguishing two genera of nematodes. One nematode has two ovaries opening into a common vulva (the primitive condition), while the other has only one ovary directed anteriorly from the vulva. The distinction between the two ovaries results from an alteration in cell fate so that the ovary primordium undergoes programmed cell death in one genus but not in the other. The finding that precisely the same alterations can be caused by a single gene mutation is evidence of a gene for the switching of developmental pathways.

At later stages of development, genetic programmes may be switched in response to environmental signals. These signals may shut off further development, as in diapause in insects (Doane, 1973), where certain seasonal changes in daylength or temperature halt development at specific stages of the life cycle. Environmental signals may also cause a switch from one developmental programme or circuit to an alternative one, particularly in insects. Examples are caste determination in Hymenoptera (Brian, 1965) and the switch between bisexual reproduction and parthenogenesis in aphids (Lees, 1966). In these instances, the expression of one set of genes is blocked while another set is turned on in competent tissues or organs at particular critical stages of development. The switch is usually in response to environmental signals which are conveyed by hormones or neurotransmitters.

One direction of evolution can be detected in these irreversible switchings of programmes of gene expression. Where two or more phenotypes are stably expressed from a common genotype, different pathways of gene action operate early in development, e.g. in the specification of insect castes. But insects that change by metamorphosis from a fully differentiated larva to a completely different adult morphology have a compartmentalised embryonic system involving a physical subdivision into populations of larval and adult cells. The cells that form the adult remain in an undifferentiated state while larval cells undergo differentiation and morphogenesis. At metamorphosis most larval structures are absorbed, and a different set of genes directs the formation of the adult structures from the undifferentiated cells (Gilbert and Frieden, 1981). A comparative study of metamorphosis in amphibian genera suggests that evolution has been towards an increasing divergence of larval and adult morphology accompanied by the appearance of increasing developmental compartmentation (Alberch, 1986).

Polarity and segmentation

Much of animal development is concerned with the specification and differentiation of metameric segments. In both the nematode (*Caenorhabditis*), which shows few signs of segmental body pattern, and the leech (*Helobdella*), which is highly metamerised, the cells – including those forming the metameric segmentation pattern – exhibit a high degree of autonomy (cell intrinsic programming) during development (Zackson, 1984). In arthropods and vertebrates, in contrast, morphogenesis of segments arises early in embryogenesis by the formation of a series of cell groups, each of which gives rise to a precisely defined part or subsegmental domain (compartment). Each segment appears to be divided into an anterior and a posterior compartment (Weisblat, 1985). Thus *Dropsophila* compartments like wing imaginal discs are made in two pieces, each of which originates from two distinct pools of cells that are established before the disc is formed.

In *Drosophila*, maternal-effect genes such as Bicaudal and Dorsal are expressed during embryogenesis where they determine the polarity of the embryo. Following the establishment of axes of polarity, the segmentation genes like Engrailed and Fushi Tarazu act before the cellular blastomere forms, to divide the embryo into segments. Others, such as the homoeotic genes like Bithorax and Antennapedia specify the form and cell pattern of particular segments. Both the segmentation and homoeotic genes appear to code for basic proteins which bind to specific sequences in the DNA, where they may act as regulators of transcription (Desplan et al., 1985). A characteristic of several of these developmental genes of *Drosophila* is that they include a conserved protein-coding region (homoeo box) of 180 to 200 nucleotides, which has been found in a wide range of higher animal phyla (McGinnis et al., 1984; Table 9). This box codes for a protein homologous with that coded by the yeast mating-type locus which regulates a number of other yeast genes determining whether the haploid cell is of a- or α- mating type (Shepherd et al., 1984). This finding, plus the fact that homoeotic genes in animals are expressed during embryogenesis, suggest that they act as regulatory genes that select between alternative developmental programmes.

TABLE 9

The phylogenetic distribution of homoeo box homology of the Antennapedia class as detected by hybridisation with labelled homoeo box DNA from *Drosophila* - in separate experiments, the Mollusca gave different results. Data abridged from McGinnis et al. (1984) and Holland and Hogan (1986).

Phylum or Subphylum	*Homoeo box homology*
Eubacteria (*Escherichia coli*)	not detected
Fungi (slime mould, yeast)	not detected
Angiospermae (flowering plant)	not detected
Coelenterata (sea anemone)	not detected
Aschelminthes (nematode)	not detected
Nemertea (ribbon worm)	one copy or poorly conserved
Platyhelminthes (tapeworm)	uncertain
Platyhelminthes (planarian, fluke)	not detected
Brachiopoda (lamp shell)	one copy or poorly conserved
Mollusca (gasteropod, squid)	not detected
Mollusca (snail, sea hare)	multiple copies
Annelida (earthworm, leech)	multiple copies
Arthropoda (insect, crustacean)	multiple copies
Echinodermata (sea urchin, starfish)	multiple copies
Urochordata (tunicate)	one or few copies
Cephalochordata (amphioxus)	multiple copies
Vertebrata (amphibian, bird, mammal)	multiple copies

Homoeosis and evolution

Homoeotic mutations occur most commonly in insects and other arthropods that are entirely composed of a series of segments. Raff and Kaufman (1983) have described how homoeotic mutations of the type known in *Drosophila* could have been primarily responsible for the evolution of contemporary insects. The proposed pathway is from the annelids (roundworms) through the Onychophora (animals with both annelid and arthropod features) to the myriapods, the primitive wingless insects and finally to the advanced winged insects.

The annelids are considered to be ancestral to insects because both have a metameric organisation, a ventral nervous system and a dorsal heart. However, in annelids the cleavages of the early embryo are spiral and the blastomeres so produced are determinate (programmed to follow a specific course of differentiation). In arthropods, in contrast, the embryo usually undergoes a series of 13 nuclear divisions before cell wall formation takes place, cell fates become determined, and the zygotic genome begins transcription. Following gastrulation the typical annelid larva consists of three discrete regions separated by bands of cilia. The middle (trunk) region becomes further divided into three segments and, as the larva grows, more segments are added from a subterminal growth zone.

This pattern of segmentation has been much altered in *Peripatus* (class Onychophora). Segments behind the mouth have been directed and elaborated into segments that give rise to the mouthparts. The change is in part due to homoeotic mutations affecting the Proboscipedia gene of the Antennapedia complex which determines the development of the second and third thoracic segments. Some alleles of this complex are known to transform palps into antennae, for they act as alternate switches controlling segmental fate. Further changes in segmentation pattern in the myriapods give six segments composing the head in place of the three in the Onychophora. The main difference between the myriapods and the wingless insects (apterygotes) in development is the division in the latter of the trunk into a three-segment thorax and an eight-segment abdomen. There is also a suppression of limbs in the segments posterior to the third trunk segment. The succeeding winged insects (pterygotes) have the three thoracic segments differentiated such that the second and third segments bear the wings. The Diptera evolved later by a reduction of the metathoracic wings to halteres, a diminution in size of the dorsal prothorax and a complete suppression of the abdominal limbs.

Many of the morphological transformations in insect evolution appear to have involved sequential additions or diversifications of loci in the homoeotic complexes. For example, the deletion of two loci (Bithorax and Postbithorax) transforms a two-winged *Drosophila* into a more primitive four-winged insect. A deletion of one of the loci of the Antennapedia complex results in three similar thoracic segments without wings as in the apterygotes. With a deletion of the Bithorax complex, a series of similar thoracic segments form, like those of certain myriapods. When both the Antennapedia and Bithorax complexes are deleted, an organism like *Peripatus* results. Raff and Kaufman (1983) regard these homoeotic complexes as ones involved in the specification of pattern in the interpretation of position during ontogeny. These complexes presumably do not supply the genes responsible for wings and legs but act as controlling switches to alter the genes from expression to nonexpression, or the reverse. Such genes may undergo alteration to produce novel morphogenic patterns which are compartmentalised in that their alternate states can both form viable links in the ontogenetic progression. For this reason, they may be of great importance in evolution, as the *Drosophila* homoeotic mutants indicate.

2.3.4 Conclusion

A characteristic of the prokaryotes is that they never achieved any but the simplest form of differentiation. Despite occasional incursions into multicellularity their cells remained practically undifferentiated, and prokaryotic organisms have not surpassed the microscopic level of existence. This primitive state has usually been attributed to the relatively poor energy transformation systems they possess. If this assumption is correct, differentiation should be largely a property of the much more energetically efficient eukaryotes. This is, in fact, shown at two levels. The first is in the lower eukaryotes, where the predatory ciliates have evolved, within a single cell, intracellular structures which are morphologically and functionally reminiscent of some of the major organs of multicellular animals. The second type of differentiation, resting on separate genetic programmes to form different types of cells, has developed weakly in plants but elaborately in animals.

The reason for the low level of plant differentiation is unclear, but it is probably connected with the necessity for differentiated cells in morphogenesis, which is but little evolved in plants. Their cellulose walls, besides preventing morphogenic cellular movement, have shut plants off from such other important morphogenic processes as cellular adhesion and surface interaction and reciprocal tissue inductions. In their absence, plants have use only for a low level of cellular differentiation which, even with a control of the planes of cell division, produces a rather stereotyped morphology and a limited set of functions.

In plants, it is apparent that many changes affecting single developmental genes give large viable modifications of the phenotype. This is assumed to be a consequence of a low level of cellular differentiation and a distribution of ontogenesis among many more-or-less independent meristems. Perhaps as a result of the simple state of plant morphogenesis, changes in development that characterise species, genera, families and orders seem to be of the same type throughout the range. The underlying genetic changes, whatever their nature, affect both major developmental genes and modifying genes, but their coincidental occurrence suggests they are caused by the same mechanism.

The situation is very different in the higher animals. Their mode of nutrition has required the evolution of various types of movement, of coordination by nervous systems, of sense organs and of organs for the intake, digestion, absorption and excretion of food and its products.

Among the animals, therefore, highly complex patterns of differentiation and morphogenesis, each with a multitude of variations, have fairly rapidly evolved. Animal ontogeny is a closely integrated interdependent process with successive stages requiring the accurate fashioning of earlier stages for their completion. The sequential nature of animal ontogenesis means that it may be easily upset by mutations of large effect, and the integration of the processes involved ensures that even minor mutations have pronounced pleiotropic effects. As a result, many viable morphological mutations only change peripheral and terminally-evolved characters such as fur colour and hair length, beak shape and plumage pattern, tooth number and structure, claw and foot type, size and, within limits, shape of the body and so on. Genetic changes that give new orders and classes of animals, in contrast, appear to be ones which can suddenly produce large alterations in morphology without seriously affecting the animal's ontogeny.

There appear to be several possibilities regarding the origins of innovations: they are due to changes in developmental switches such as give one or two ovaries in the nematode; they are alterations which affect coordinations, as in the formation of sensory structures in conjunction with the endocrine system; or, as the conventional view has it, they come about solely through the accumulation of mutations that have slight effects at any developmental stage. The matter is still under debate, but attention has been drawn to the probability that, in animals, genetic changes are likely to be tolerated only if they occur at stages of development when inductive and other types of interactive events are not occurring. At these stages, a certain degree of dissociation of developmental parts is possible, permitting mosaic evolution, heterochrony etc.

From the types of developmental changes associated with the origin of various animal taxa, we might expect to gain some impression as to which processes are important at different evolutionary levels. But the data are still so few that all we can say is that, at the level of order and above, embryonic patterns of determination and morphogenic movements have remained much the same since their origin. Nor have programmes for cell differentiation altered very much, although the arrangement of cells in tissue may differ as between orders. This is presumably due to differences in induction-reaction processes, in particular to changes in reactive responses to inductive stimuli. A less fundamental group of developmental processes seems to change more frequently. It includes neoteny, which is believed to have been responsible, for example, for the origin of some flightless birds, allometry, which gives variation in shape with change in size, and programmes of cell death where alterations affect the degree of webbing between digits etc. Since thyroxine is involved in amphibian neoteny and chalones (specific mitotic inhibitors) are thought to determine organ size, some of these developmental changes may be hormonal in origin. As we found in considering the effects of selection in animal breeding, alterations in hormonal systems seem to provide a frequent cause of change in body shape and composition. These effects are exerted through changes in one of the *trans*-acting systems of developmental regulation. They are, however, accompanied by a high level of direct pleiotropy, so, in the formation of the lower taxa, it is more common for the individual *cis*-acting components of gene regulation to be altered.

A final possible cause in the formation of orders and classes rests on the proposition that gross changes in morphological and physiological characters frequently require the origin of new genes. The possibility of new gene formation in turn depends upon the structure of existing genes and the ability of the genotype to recombine them to provide new ones. Such data as there are supporting this proposal are considered in Section 4.5, which deals with evolutionary assessments of phyla and classes.

PART 3
Mechanisms of Genetic Change

Since Darwin's time, it has been generally accepted that evolution is driven by the concerted action of two sequential processes. The first is the continual occurrence of genetic changes generated by random mutation and the second is selection for adaptively advantageous phenotypes and selection against those that are disadvantageous.

In its broadest sense, the term mutation encompasses any inherited modifications of the genotype that may affect the phenotype. Mutations are conventionally subdivided into point mutations and chromosomal alterations, but these terms were given before the nature of mutations at the DNA level became known. Chromosomal changes are cytologically detectable, while hereditary changes that were not were classed as point mutations. For our purposes we will take point mutations to comprise the replacement, deletion or addition of one or a few nucleotides within a gene. This definition is necessary to distinguish point mutations from other less familiar types of genetic change, called nonmutational, which have a different origin. Point mutations are the result of diverse causes, including nucleotide tautomerism, alkylation, dimer formation, the inexactitude of some DNA repair processes and additions or deletions of nucleotides arising from recombinational errors. The other component of mutation, as the term is used here, is chromosomal change but, as discussed above (Section 1.4), there is little evidence for an evolutionary role for chromosomal mutations other than in the occasional production of positional effects and perhaps in reproductive isolation in species formation.

There can be little doubt that point mutations must have been very important in establishing the early stages of life. But thereafter there would seem to be selective advantages to supplementing the primary mutation mechanism (nucleotide substitution) with others that were less random in origin, more frequent in occurrence and less deleterious in phenotypic effect. The following pages examine the proposition that

96

mechanisms providing genetic variation for evolution have themselves evolved; other components of the evolutionary process such as reproduction, inheritance and competition are not considered.

In assessing the nature of additional (nonmutational) mechanisms, we must recognise that mechanisms cannot be evolved to deal with future contingencies, but must have an immediate selective advantage which is independent of the mechanism. A simple example is the selection for a mutant where the mutation was induced by a mutator gene. A frequent result will be selection, not only for the advantageous mutation, but also for the mechanisms that produced it; i.e. the mutator gene.

3.1 *Role of Mutation in Evolution*

For the first few hundred million years after the origin of life, spontaneous mutation was probably the main source of transmissible genetic change for the slow building-up of single-celled life forms. For the genesis of enzymes, for example, numerous replacements of amino acids in simple, relatively unordered polypeptides must have occurred before the basic six or so types of bond rearrangements in catalysis had been established (Langridge, 1982). Then the replacement of these few enzymes, weak in catalytic activity and low in specificity, by many descendent ones with progressively increasing specificity and catalytic efficiency would require many more mutational substitutions as well as template multiplication. Mutation must also have been essential in the development of enzyme-controlling systems, membranes and other structural proteins. All this foundational evolution of genes and proteins must have occurred in the absence of the recombinational advantages of sexuality and the other special mechanisms discussed below, which now accelerate evolution.

3.1.1 *Rates of spontaneous mutation*

Mutation rates in primeval organisms or molecules were likely to have been much higher than those of today, for two reasons. Until about 2500 million years ago, when oxygen began to be produced on earth (Olson and Pierson, 1986), initiating a shielding layer of ozone in the lower stratosphere, the flux of highly mutagenic ultraviolet light was very intense. Secondly, the initial replicating nucleic acid forming the genetic basis of life is thought to have been of RNA because DNA molecules are synthesised via RNA intermediates and RNA primers are required for DNA replication. None of the RNA polymerases has been shown to have an error-correcting exonuclease associated with it, as is found with DNA polymerases. Consequently, the modern RNA genomes of some viruses have spontaneous mutation rates which are about a million-fold higher than those of similar DNA ones. But if the first replicating molecules were indeed of RNA they were later replaced by DNA ones of greater stability. The half-life for the hydrolysis of phosphodiester bonds is about 360 years in DNA, but it is only 30 years in RNA (Miller and Orgel, 1974). The change in genetic material from one form of nucleic acid (RNA) to another (DNA) may not have been particularly difficult; Temin and Mizutani (1970) have found that DNA can be synthesised from an RNA template.

Most *in vitro* measurements of error rate have been based on the frequency of incorporation of noncomplementary nucleotides during replication with synthetic polynucleotide chains as templates. They have shown that RNA polymerase gives an error rate for RNA replication of about 2×10^{-3} per incorporated nucleotide, which probably represents the frequency of unusual tautomer formation. In contrast, the DNA polymerase of *Escherichia* with DNA as a template has an *in vitro* error rate of only 10^{-6} to 10^{-7} (Table 10). The greatly lowered rate of misincorporation with the bacterial DNA polymerase is believed to be due to its associated corrective function.

TABLE 10

Representative estimates of misincorporation or mutation frequency per base-pair per replication.

	RNA	Viral DNA	Prokaryotic DNA	Eukaryotic DNA
In vitro	4×10^{-3} – 2×10^{-4} (1)	10^{-5} – 10^{-6} (3)	10^{-6} – 10^{-7} (5)	5×10^{-5} (7)
In vivo	10^{-3} – 10^{-4} (2)	10^{-8} (4)	5×10^{-9} – 5×10^{-10} (6)	5×10^{-9} (8) – 5×10^{-11} (9)

(1) Site-specific mutation of Qß RNA   Batschelet et al., 1976
(2) Qß RNA synthesis with RNA polymerase   Domingo et al., 1978
(3) DNA synthesis by T4 replicating complex   Hilner and Alberts, 1980
(4) Mutation frequency in T4 and λ   Drake, 1969
(5) DNA synthesis by *E. coli* polymerase I   Kunkel and Loeb, 1980
(6) Reversion frequency in *E. coli* tryptophan   Fowler et al., 1974
    gene
(7) DNA synthesis by mammalian polymerase α   Kunkel and Loeb, 1981
(8) Mutation frequency in mammalian α-globin   Li et al., 1981
    pseudogenes
(9) Mutation frequency in *Drosophila*   Bernardi and Ninio, 1978

Errors that arise during *in vitro* synthesis are usually measured by mutation frequency, often by the reversion to normal of a mutation of known molecular nature that has inactivated a function of the organism. Using such tests, it is found that with *in vivo* replication, the rate of misincorporation for RNA is little changed, whereas that for prokaryotic

DNA is reduced below the *in vitro* level to 5×10^{-9} to 5×10^{-10} per nucleotide pair per replication (Table 10). The DNA error level leading to reverse mutation in prokaryotes is reduced by base selection where the replication enzymes discriminate against incorrect nucleotides, proof-reading where corrections of misincorporation are made during the replication process, and mismatch repair where there is a post-synthetic correction of nucleotide mismatches (Loeb and Kunkel, 1982). In eukaryotes, where the DNA polymerases themselves are less accurate, additional mechanisms, perhaps other enzymes and DNA-binding proteins, appear to operate to ensure fidelity of replication. It seems likely that the spontaneous DNA mutation rate, and consequently its contribution to evolution, has been severely reduced by selection against base misincorporation which has probably been brought to the minimum level compatible with the biochemical cost of error correction.

3.1.2 *Dampening of mutational expression*
Despite the presence of effective repair mechanisms, more genetic variation or genetic damage may occur naturally than species could tolerate, unless means of dampening its expression were available. Such dampening may be found in the organisation of the code (Appendix 1), which appears to be constructed so as to minimise the phenotypic effects of single nucleotide substitutions.

A large fraction of nucleotide change is absorbed by degeneracy, the presence in the code of more than one triplet for a particular amino acid. In a code where 64 triplet combinations of four types of nucleotide code for only 20 amino acids, degeneracy prevents frequent mutation to noncoding triplets which would cause a break in the polypeptide chain during translation. The three noncoding RNA triplets left in microorganisms (UAG, UAA, UGA) mean that only about 4% of substitutions lead to noncoding. Also, a series of related triplets for the same amino acid prevents many of the nucleotide alterations from simultaneously altering the amino acid sequence of the protein. Of the possible 576 (64 x 3 x 3) single substitutions of all bases in the code, nearly one quarter (134) are synonymous; that is, no change in amino acid follows the nucleotide change.

An additional feature of the code arrangement is that for many changes the new triplet specifies an amino acid whose substitution in the protein does not affect its function. In particular, amino acids with similar side-chain properties are coded for by triplets of related nucleotide composition. Volkenstein (1965) has established that polar (hydrophilic) amino acids are mainly coded by triplets high in adenine and cytosine, while nonpolar (hydrophobic) amino acids tend to be determined by triplets high in guanine and uracil. In extending this relationship, Goldberg and Wittes (1966) have observed that the frequency of mutational change from polar to nonpolar property, and the reverse, is correlated with the degree of nonpolar character that the particular amino acid exhibits. For example, nonpolar amino acids, with the exception of tyrosine, always mutate by transition to another nonpolar amino acid. In accordance with expectation, it is found that the most common naturally occurring amino acid

99

replacements caused by single-step mutations in evolution involve rather small differences in side-chain character (Zuckerkandl and Pauling, 1965; Clarke, 1970).

The importance of this dampening of the effects of mutational substitution is increased by the asymmetrical distribution of amino acids in soluble proteins. In proteins of known conformation, nearly all polar and ionic side-chains are in contact with water either at the surface of the molecule or in cavities. These amino acids may frequently be interchanged without affecting protein function. Changes that still give a functional molecule of haemoglobin (Perutz and Lehmann, 1968) or cytochrome C (Dickerson, 1971) are mostly replacements of one type of polar side-chain by another polar one at the surface, but even certain changes to nonpolar side-chains are tolerated. Large nonpolar side-chains lie in the interior of the molecule, in crevices so designed as to minimise contact of these side-chains with water, or on the surface where they form points of contact between subunits. The substitution of an interior nonpolar amino acid for a polar one would destroy enzymatic activity, but the code structure minimises such changes (Figure 20).

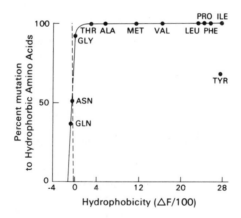

Figure 20 Diagram of the frequency with which codons for hydrophobic amino acids mutate to specify other hydrophobic amino acids. The free energy change (ΔF) when an amino acid is transferred from a nonpolar solvent to water is plotted on the abscissa. Hydrophobic amino acids are ones in which the free energy change exceeds that of glycine. The data show that the codon structure effectively prevents a change of hydrophobic amino acids to ones with a hydrophilic character following transition, the most frequent form of nucleotide substitution (figure abridged from Goldberg and Wittes, 1966).

Thus there are at least three structural features at the DNA and protein level which tend to buffer the organism against mutational change: multiple codons for most amino acids; correlations between coding triplets and side-chain character; and restriction of polar amino acids to the surfaces of soluble proteins.

3.1.3 *A test of buffering efficiency*

This test was made by determining the relative frequency of unexpressed mutational changes in several genes of the lactose operon of *Escherichia coli* (Langridge, 1974). From the genetic code and the amino acid composition of a protein, we can calculate the expected occurrence of chain-terminating (noncoding) and of missense mutants (amino acid replacements) following nucleotide substitution. Excluding nucleotide changes which do not alter the amino acid sequence, we would expect 6% of mutations in the ß-galactosidase gene to be chain-terminating and 94% to be missense. But in fact, of those mutations reducing enzyme activity to less than 50% of the normal level, not 6% but 72% are chain-terminating. Most amino acid replacements appear to have little effect on enzyme activity and those few that do probably occur at or near the active sites of the enzyme. On the other hand, nearly one-third of amino acid replacements cause inactivation of the galactoside permease. This protein may be more sensitive than ß-galactosidase to mutation because, in addition to catalysing sugar transport, it requires a precise conformation for fitting into membrane sites. Of mutations affecting the lactose operon repressor, only 3% are chain-terminating. This frequency is close to that expected if all nucleotide substitutions that are expressed in the protein are phenotypically detectable. The data suggest that sensitivity to mutation is greatest in genes involved in nucleic acid - protein interactions (e.g. operator, repressor), less so in genes whose products form part of a structural complex (e.g. permease) and least in genes specifying metabolic enzymes (e.g. ß-galactosidase).

The strict nucleic acid sequence requirement for protein-binding sites is also shown by promoters which initiate transcription, but do not code for proteins. For example, in the mouse ß-globin promoter, over 80% of the nucleotides are inert to mutational substitution, but any mutation in the three regions to which specific transcription factors attach reduces or very occasionally increases transcription (Myers et al., 1986).

There is good reason to believe that, since the origin of self-replicating nucleic acid molecules, the assumed primary driving force of evolution, mutation, has greatly decreased in both frequency and effect. The probable change in template composition, the construction of the genetic code, the evolution of error-correcting systems and the asymmetrical structure of enzymatic proteins may all have occurred because mutation, in the main, is more damaging than beneficial to the organism.

3.1.4 *Tests of evolutionary effectiveness of point mutation*

Although mutation may not now be the evolutionary force that it once was, it might still be sufficient in frequency and scope to account for much of the evolution that has occurred during, say, the last few hundred million years. Information on this matter is provided by the study of artificial evolution in bacteria, the only organisms suitable for such experiments. For example, bacterial numbers of 10^{12}, mutagens that increase mutation frequency 10^3- to 10^6-fold and selection pressures many times in excess of those of natural selection can be used in a single experiment. Since bacteria are haploid, it is relatively easy experimentally to attain the equivalent of the tens of millions of years of evolution of comparable genes in many higher eukaryotes.

TABLE 11

Michaelis constants (Km) or competitive inhibition constants (Ki) in molar concentrations, and maximum velocities (V) in moles per 30 minutes for the hydrolysis of glycosides used in evolutionary experiments with *Escherichia coli*. The kinetic constants for the hydrolysis of the natural substrate of ß-galactosidase, lactose, are included for comparison. Only mutations in the ß-galactosidase gene allowing hydrolysis of lactobionic acid were obtained.

	Km or Ki	V
4-(ß-D-galactosido)-D-glucose (lactose)	1.5×10^{-3}	5.5×10^{-9}
4-ß-D-galactosido)-D-gluconic acid (lactobionic acid)	6.5×10^{-2}	1.0×10^{-9}
Phenyl-ß-D-thiogalactoside	3×10^{-3}	0
6-(a-D-galactosido)-D-glucose (melibiose)	4×10^{-2}	0
Methyl-ß-D-galacturonic acid	0	0
Methyl-ß-D-glucose	0	0

However, the extensive experiments reviewed by Clarke (1980) and Mortlock (1983) have demonstrated only a recurrent, limited response of the bacterial genome to the presentation of new sources of carbon or energy, or to the presence of toxins. In general, the organisms acquire through mutation the ability to use a new carbon source by inactivation of induction or control systems or by minor shifts in substrate binding, and to resist toxins by alterations in permeability or changes in ribosomes

(Section 2.1.1). No new gene has appeared in response to the intense selection, no alteration in bond rearranging mechanisms has been observed in any enzyme and no new catalytic specifications have been found.

As a further test of the potential ability of mutation for evolving new genes, a series of experiments was performed (Langridge, 1969 and unpublished) in which selection was exerted for the hydrolysis of various derivatives of lactose by mutant ß-galactosidase in *Escherichia coli* (Table 11). The compounds are not hydrolysed either because they bind poorly to the enzyme or because the glycosidic bond is incorrectly oriented for the enzyme's catalytic amino acids. The changes expected in the enzyme for efficient hydrolysis are as follows: for lactobionic acid, an increase in the hydrogen bonding or van de Waals forces to the gluconic acid part; for the thiogalactoside, a hydrogen-bond type of linkage between two sulphur atoms; for melibiose, a change in the relative positions of the catalytic sulphydryl and imidazolium side chains; for the galacturonate, a reduction in steric hindrance; for the glucoside, the formation of a new hydrogen bond in the enzyme-carbohydrate complex. However, mutations were obtained only for the hydrolysis of lactose with an open glucose ring (lactobionic acid). These occurred either by constitutive ß-galactosidase synthesis plus multiplication of the corresponding gene, or by alterations in substrate affinity. No mutants were found in which the enzyme could split an α- in place of a ß-linkage, the S- in place of the O- glycoside, the substitution of -COOH for $-CH_2OH$ at the 6'- position or the inversion of the 4'- hydroxyl in the galactose moiety.

Although large numbers of the survivors of mutagenesis with different mutagens in repeated experiments were examined, the results were the same as are usually found; i.e. mutation of the relevant gene can produce adaptive changes to only very small alterations in substrate structure. The mutations calculated to have occurred in these experiments would have included all possible single transitions as well as most transversions, and selective conditions were such that even a trace of new catalytic activity could be detected. Although similar experiments have not been done with higher organisms because of their unsuitability (diploidy, long generation time, large size etc.), it would be anticipated that their more complex genes would not be any more amenable than those in bacteria to the influence of point mutation.

We may conclude that, while nucleotide substitution was once sufficient to evolve new enzymes with entirely new specificities and catalytic mechanisms, it may no longer be able to do so. Not only is it inadequate in this respect, but it is now a degenerative process because, as organisms become more complex it is more and more likely that simple mutational events will produce deleterious or inactivating genetic changes that are selectively disadvantageous or lethal. Because spontaneous mutation is now an unavoidable liability to the organism, it has been reduced to a minimum in frequency and the intensity of its effects has been diminished as much as possible.

This conclusion raises two questions. The first is whether new genes and enzymes were only made soon after life arose when mutations were more frequent and more readily expressed, or have genes and enzymes evolved in complexity? When we reflect that the probability of forming the sequence of a relatively small gene of 1000 nucleotides by random polymerisation is about 10^{-600}, it is obvious that new genes cannot arise by such a means. A reasonable alternative is to suppose that genes and enzymes were initially radically simpler than they are now and, as they have increased in specificity and catalytic efficiency, they have become much larger. An example perhaps is the esterase of five amino acids synthesised by Langridge and Bentley (unpublished). While histidine alone gave a weak stoichiometric catalysis of a chromogenic ester, the synthetic pentapeptide had the typical enzymatic action of substrate binding followed by hydrolysis. One way for such a catalytic peptide and its coding gene to increase in size is by tandem duplication, as seen in the ferredoxin molecule of *Clostridium* (Eck and Dayhoff, 1966). This protein, which was one of the earliest electron carriers in anaerobic bacteria, consists of two very similar sequences of 26 amino acids which have been joined together. The original gene was linearly duplicated, and a fused protein resulted from the conversion into a coding triplet of a sequence which formerly terminated the polypeptide chain (Figure 21).

```
Ala Tyr Lys Ile | Ala Asp | Ser | Cys | Val Ser | Cys Gly | Ala | Cys Ala | Ser Glu | Cys Pro Val | Asn | Ala | Ile Ser | Gln | Gly Asp Ser Ile
 1   2   3   4  |  5   6  |  7  |  8  |  9   10 | 11  12  | 13  | 14  15  | 16  17  | 18  19  20  | 21  | 22  | 23  24  | 25  | 26  27  28  29

Phe Val Ile Asp | Ala Asp | Thr | Cys | Ile Asp | Cys Gly | Asn | Cys Ala | Asn Val | Cys Pro Val | Gly | Ala | Pro Val | Gln | Glu
30  31  32  33  | 34  35  | 36  | 37  | 38  39  | 40  41  | 42  | 43  44  | 45  46  | 47  48  49  | 50  | 51  | 52  53  | 54  | 55
```

Figure 21 Internal duplication in the sequence of ferredoxin from *Clostridium* (Epstein, 1970). The amino acids from 1 to 26 are homologous with those in the sequence 30 to 55.

The second question is whether organisms have ceased the evolution of new genes and enzymes except for making minor adaptive alterations, or whether they have developed mechanisms for such evolution other than point mutation. Since major evolutionary change has clearly occurred in many organisms even during the last few hundred million years, it seems that other mechanisms of evolving must be available to them. We may find what these mechanisms are by considering the nature of gene evolution in viruses, prokaryotes and eukaryotes. Our information on this matter comes in part from mechanisms for the production of genetic variation and in part from mechanisms of evolutionary change that are known to have occurred.

3.2 *Viral Evolution*
The size of the genome in different viruses varies enormously, as do the opportunities for evolutionary change, from ones containing only two or

three genes for replication and a coat protein to others having 200 to 300 genes. In the larger viruses, proportionately more of the genome is usually directed to stopping host metabolism or subverting it towards the production of virus.

It might be expected that viruses would use the same mechanisms for evolution as their hosts because they are largely dependent on the enzymes of the host cell. However, viruses differ importantly from free-living organisms in respect to the small size of the genomes, often in the nature of their genetic material and in the very large numbers generated during replication; these differences may have conditioned a variety of evolutionary strategies.

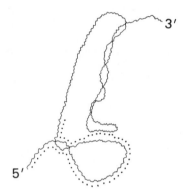

Figure 22 A model of discontinuous transcription of RNA viruses. The transcriptional jumping of RNA polymerase is thought to occur within a replication complex.

In viruses having RNA genomes, mutation may remain the most important factor in their evolution. Mutation rates by nucleotide substitution are very high, sometimes exceeding 10% per genome per replication (Holland et al., 1982). Such high rates appear to be due to the absence of error-correcting nucleases, for none has been found associated with RNA polymerases. Although the genomes are small, probably because a large genome would be too sensitive to such high mutation rates, the rates may in fact be higher than the forces of natural selection can completely cope with. Domingo et al. (1978), after examining the nucleotide sequence heterogeneity in a population of Qβ RNA bacteriophage, concluded that the virus did not have a defined unique structure but consisted of a large proportion of slightly different individual sequences. They calculated that the mutation rate is such that an RNA bacteriophage population of 10^{11} members would contain all possible single, double and triple transition mutants.   However, point mutation may be an evolutionary mechanism conferring only minor changes in these viruses, with major ones being the result of intermolecular genetic rearrangement. Viroids (Keese and Symons, 1985), plant viruses (Haseloff et al., 1984) and animal RNA viruses (McCahon and Slade, 1981) have all been reported to exchange sequences between molecules or between domains of a molecule, possibly via discontinuous transcription on one template and re-initiation on another (Jennings et al., 1983;   Figure 22).   The model of exchange envisages regions of sequence brought into proximity by the tertiary structure of a 'replication complex' so that the viral polymerase can shift between adjacent regions on the RNA template during replicative synthesis.

RNA viruses with segmented genomes, that is, with genomes divided into two or more molecules, are found in plants and animals, but not in bacteria. They not only have high mutation rates but also can evolve by segmental reassortment in doubly infected cells and perhaps by true recombination. Both of these processes may help to reduce the lethal effects of mutation by creating viable genomes out of damaged ones, whereas the unsegmented RNA viruses must rely on their small genome size, the large numbers produced on infection and possibly discontinuous transcription.

The main evolutionary changes which occur in RNA viruses now concern virulence.   In this connection, Webster et al. (1982) have shown that in human influenza virus, antigenic shift, a complete change in antigenic properties, appears to occur by reassortment of segments, while antigenic drift is due to small changes in antigenic specificity resulting from the gradual accumulation of point mutations.   For example, in the Hong Kong strain of human influenza, a strain resulting from antigenic shift, the entire gene for the surface antigen, haemagglutinin, has apparently been replaced by one from a duck or horse virus. The subsequent antigenic drift of the Hong Kong strain was brought about by 2 to 33 amino acid changes, according to variant, in the antigenic sites of haemagglutinin.   Virulence changes particularly affect haemagglutinin because this protein is responsible for the attachment of virus to cell.

The variation in the RNA viruses is such that, in the view of Nahmias and Reanney (1977), the conserved aspect of a viral genome is its basic genetic organisation - the grouping, spacing and order of genes - rather than the encoded sequences as such.

In the bacterial DNA viruses, spontaneous mutation rates are similar to those in bacteria when expressed per genome (Drake, 1969).   The best studied of these viruses, (λ, P 22, phage 21 and phage 80) appear to evolve together as a group by recombining genetic segments which have independently evolved for particular viral functions.   In these genomes, certain genes occur in blocks (modules) containing one or more genes for a particular biological function.   Although the modules may vary considerably as mutation inevitably occurs, the nucleotide sequences between them are kept invariable so as to provide sequence homology for recombination of the modules (Botstein, 1980).   Thus, in addition to selection in recombinants for functional efficiency, there is also selection for the maintenance of the recombination system (the homologous

106

sequences). The product of evolution is not a particular virus but a family of genetic modules, and further evolution takes place by module exchange (homologous recombination), not by linear descent (Figure 23). In addition to these lambdoid bacteriophages, a similar evolutionary mechanism seems to be operative in hetero-immune viruses of *Bacillus licheniformis* (Storchová et al., 1985). The genomes of these bacteriophages diverge from each other in the substitution, insertion or deletion of whole DNA segments rather than by point mutations. This is in accord with the modular model, as is also the fact that recombination is not detectable between genes coding for structural proteins, which have coevolved and must cooperate in viral head assembly.

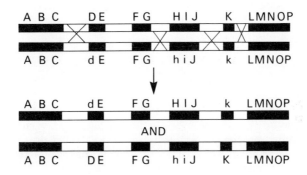

Figure 23 A model of modular recombination in bacteriophage. The letters are gene symbols.

In many animal viruses, especially the DNA tumour viruses and the RNA retroviruses, additional variation is generated, not by modular exchange but by differential RNA splicing (Ziff, 1985; Figure 24). The fixed size of most viral genomes has required evolution to maximise the versatility of template-encoded information by making a variety of different messenger RNAs from a primary RNA transcript. The pattern of RNA splicing sometimes appears to follow a programme because structurally novel proteins are produced at specific stages of the viral life-cycle. However, it is not clear whether the splicing mechanisms and their control were evolved by the viruses or whether they merely used the mechanisms of the host cell. Nor is the nature of the programme clear; it may be genetically determined but the products of differential splicing are not inherited, except occasionally in retroviruses by reverse transcriptional integration into the nuclear genome.

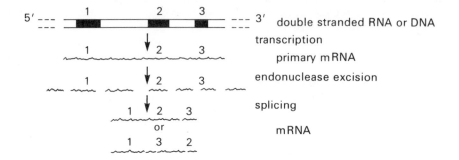

Figure 24 Diagram of normal and abnormal RNA splicing. Numbered segments are coding sequences within a gene; unnumbered segments are intervening sequences.

We see that, even in genomes as simple as those of viruses, there is a variety of mechanisms for genetic change apart from point mutation. A mechanism like segmental reassortment seems analogous to chromosomal segregation, but others like discontinuous transcription and modular exchange are unique to some viruses, as far as we know. Although the evidence is limited, it suggests that for certain viruses, these mechanisms may be more important in their evolution than point mutation.

3.3 *Prokaryotic Evolution*
We have already established that mutation is relatively ineffective in prokaryotes for the generation of evolutionary novelty, yet bacterial evolution is observed frequently to occur today. The main means by which these and other adaptive changes in the genome are brought about have been widely known since a review of the Japanese literature by Watanabe (1963), and they are quite unlike those used by other organisms.

3.3.1 *Plasmids and transposons*
In both bacteria and cyanobacteria, the genome is usually in two parts, chromosome and plasmid, which seem to separate genes according to their evolutionary significance. While the chromosome contains the essential genes for growth, reproduction and metabolism, the plasmid has genes dealing with particular factors of the environment such as poisons, competing organisms or uncommon carbon and energy sources. There is often a collection of genes on the one plasmid providing information to deal with a range of antibiotics, or sequential steps to convert an organic compound into one of the usual metabolic intermediates of the cell. Therefore, the complete genetic information to adapt to a new ecological

condition can be acquired in a single step. Most plasmids are transmissible to other cells by conjugation, a sexual process involving a one-way transfer of genes from one mating type to another, which is common in bacteria; for example 30 to 50% of *Escherichia coli* strains isolated from nature possess conjugal plasmids (Curtis et al., 1970). Since plasmids can replicate, they do not need to be integrated into the host DNA for their multiplication as do chromosomal genes. Hence, plasmid-carried genes do not require sequences in common with their hosts and so can be permanently transferred to distantly related species. Wide distribution of plasmids may take place through the infection of bridging species, and alternation in gene transfer by the process of conjugation, transduction and even transformation. Conjugation is limited to gene exchange between closely related species that must have compatible proteins on their cell surfaces; transformation has frequently been shown to occur between species and even genera; and a given virus may be capable of infecting and transducing different strains and sometimes different species, but rarely different genera or higher groups. The plasmid with its included transposons and insertion sequences, which serve as portable and insertable recombination targets, thus represents a system that allows for the transfer of certain prokaryotic genes at high frequency, even from unrelated genomes. Unlike the other mechanisms, the plasmid-transposon system is not a mechanism for the production of genetic variation, but rather a means for making adaptive genetic changes.

3.3.2 *Sequence substitution*
In addition to this system for the acquisition of whole genes and gene complexes, an examination of prokaryotic gene structure indicates the presence of other nonmutational evolutionary mechanisms. One of these concerns the production of new genes, apparently by the assembly of parts of other genes. The genes for enzymes carrying out sequential but catalytically diverse reactions in the ß-ketoadipate pathway in *Acinetobacter* and *Pseudomonas* seem to have been made by a process of oligonucleotide substitution from one gene to another (Yeh and Ornston, 1980). As judged from the amino acid sequences of the enzymes concerned,

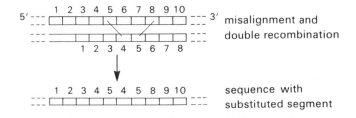

Figure 25 A possible model of oligonucleotide substitution within genes. Figures are subgenic segments.

109

relatively short regions of sequence homology have been transferred from one set of structural genes to another set. Similarly, the assembly of some *Bacillus* spore proteins has been by rearrangements of blocks of primary amino acid sequence; for example, from the carboxy-terminal and central regions of the protein of *Bacillus megaterium* to the amino-terminal regions of the proteins of *B. cerus* and *B. subtilis* (Yuan et al., 1981). Since there is no change in gene length, the DNA sequences are inserted by substitution, not addition, which apparently occurs without the usual requirement for sequence homology (Figure 25). A related evolutionary device in prokaryotes is the systems of promoter change by DNA segment inversion which serve such functions as flagella variation in *Salmonella typhimurium* (Zieg et al., 1977) and change in the host range of bacteriophage *Mu* (Kamp et al., 1978; Figure 26). In *Salmonella*, for example, the invertable DNA sequence contains a promoter which in one orientation allows the expression of genes in adjacent DNA and in the other orientation does not. Three of these inversion systems and another one catalysing deletions have been shown to contain homologous segments, the conclusion being that the invertable or deletable section has been incorporated as a module into different complex genetic structures (Plasterk et al., 1983).

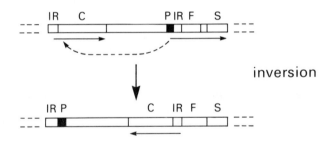

Figure 26 A diagram of promoter change by segment inversion. IR is inverted repeat; P is promoter; C is a gene catalysing the inversion; F is the gene for H2 flagellin; S is a suppressor gene for H1 flagellin; the arrows mark the direction of transcription.

3.3.3 *Gene duplication*

Although strategies for gene acquisition or evolution, such as the above, lessen the importance of nucleotide substitution in prokaryotic evolution, point mutations have frequently been invoked for the diversification of duplicated genes for different but related functions. Spontaneous gene duplication commonly occurs in prokaryotes and their viruses; for example,

110

tandem duplications arise in a single lytic cycle of λ-type viruses with a frequency of 10^{-3} to 10^{-5} (Emmons and Thomas, 1975) and duplications of specific genes may be carried by 10% of a *Salmonella* population (Miller and Roth, 1971). The mere duplication of a gene may, by increasing the level of a relevant enzyme, provide an adequate resistance to certain inhibitors, as in the duplication of the gene for ß-lactamase in *Escherichia* (Edlund et al., 1979) and the multiple duplication of plasmids or the genes in them in attaining antibiotic resistance (Section 2.1.1). Also, one member of a gene duplication may provide the substrate for a new or altered function. However, there seem to be few certain examples of divergence to a changed function solely by point mutation following gene duplication. In the known cases of a common evolutionary origin for a pair of genes, such as the repressor genes for the galactose and lactose operons of *Escherichia* (Wilcken-Bergmann and Müller-Hill, 1982) or the eukaryotic examples quoted by Ohno (1970) in his book 'Evolution by Gene Duplication', the possibility of subgenic substitution or addition cannot be excluded. The acquisition of altered function by point mutation following duplication probably occurs, but it is usually difficult to prove. Perhaps the best examples are found in the neurohypophyseal hormones of animals where amino acid substitutions are few, and physiological changes large (Section 2.1.2).

3.3.4 *Point mutation*
Even examples of adaptive change in single genes by mutation are rather rare. In prokaryotes they are practically confined to instances of slight shifts in enzyme affinity for substrates (e.g. Wu et al., 1968) and changes in ribosomal proteins conferring resistance to antibiotics (Nomura and Morgan, 1977). In this connection, it should be mentioned that presumed point mutations can be important in organisms in providing a first defence against the potentially lethal effects of antibiotics and toxins. Spontaneous mutation of genes in bacterial 'chromosomes' can alter metabolic enzymes or the proteins and RNAs of ribosomes, which are the targets of many inactivating compounds, and they can change membrane proteins to reduce or abolish the permeation of inhibitors. Such mutations are, however, usually only temporary expedients, because in the absence of the toxin, the resistant organisms are frequently at a disadvantage in competition with those that have remained susceptible (Kiser et al., 1969) In bacteria appropriate plasmid acquisition provides a less deleterious solution to the problem, with relevant transposon movement from plasmid to chromosome conferring a stable resistance.

It appears that the evolutionary requirements of contemporary prokaryotes are met mainly by mutational adjustments of existing metabolism and by the possession of horizontal transfer processes for obtaining metabolic genes as the need arises. The use of the strategy of acquiring accessory genes does not obscure phylogenetic relationships because phylogenies are deduced from sequence comparisons of ribosomal RNA, ferredoxin and cytochrome C, all of which are part of the basic and relatively stable chromosomal genome.

3.4 *Eukaryotic Evolution*

While prokaryotes possess complex evolutionary systems for transferring and establishing genes drawn from a large common pool, such horizontal gene transfer in eukaryotes is rare, if it occurs at all. Consequently, eukaryotic genes must be evolved and altered within a very restricted developmental stage, the germ line of the individual animal or the reproductive meristem of the plant. For spontaneous random genetic change, mutation is of course the acknowledged agent, but mutation rates in eukaryotes are as low, or even lower, than those in prokaryotes. As in viruses and bacteria, single nucleotide substitutions may confer antibiotic resistance (Kearsey and Craig, 1981) and altered antigen recognition (Brown et al., 1986), but examples are few. Since some eukaryotes are among the most rapidly evolving organisms, evolutionary mechanisms must be present to compensate for the feebleness of mutation and the absence of gene acquisition by plasmid- and viral-like vectors.

3.4.1 *Role of intervening sequences*

So far, four types of nucleic acid sequence change, in addition to mutation, have been recognised, although not all of them have been shown to have a role in adaptive evolution. A mechanism that does seem to be evolutionarily important derives from the flexibility given to genes by the presence of intervening sequences (introns). They are sequences within genes that do not code for protein and are widely distributed in eukaryotes, especially in vertebrates and perhaps plants (Table 12) where their presence increases the size of the average gene 10- to 30-fold (Gilbert, 1985). Following transcription, the intervening sequences are excised and degraded, while the remaining messenger RNA is ligated to form the template for protein synthesis. The intervening sequences generally separate regions of DNA which encode specific aspects of protein structure such as α-helical and ß-sheet elements, catalytic sites, export- and membrane-binding sequences and effector-binding regions (Craik et al., 1983). At present, four types of role have been detected for intervening sequences, three of them of evolutionary importance and one concerned with differentiation.

In those proteins which have a repetitive structure, the DNA coding sequence for each unit of the repeat is often separated from the adjacent coding sequence by an intervening sequence. Thus the basic unit of the chicken collagen molecule is an 18 amino acid sequence encoding a six-fold repeat of the sequence gly-x-y. These sequences interact with identical protein chains to form a triple helical structure in fibrillar collagens which is strictly conserved, probably because of the necessity for a specified length for proteins with a structural function. The 54 base-pair DNA sequence, composed of six repeats of nine base-pairs each, which codes for the collagen repeat, is believed to have been amplified (multiply duplicated) to create linear copies separated by intervening sequences (Boedtker and Aho, 1984; Figure 27). The reason for this conclusion is that, even where a coding sequence is different, it is still a multiple of nine

TABLE 12

The frequency of intervening sequences in some homologous genes from different organisms.

Gene	Vertebrate	Invertebrate	Fungus	Protist	Plant
α-tubulin	3 (man), 3 (rat)	3 (Drosophila)	1 (Schizosaccharomyces)	3 (Volvox), 0 (Stylonychia), 0 (Trypanosoma), 7 (Physarum), 2 (Chlamydomonas)	4 (Arabidopsis)
β-tubulin	3 (man), 3 (chicken)	3 (Drosophila)	5–8 (Aspergillus), 6 (Neurospora), 2 (Candida), 5 (Schizosaccharomyces), 0 (Saccharomyces)	3 (Chlamydomonas), 0 (Tetrahymena), 3 (Volvox)	2 (Arabidopsis)
Myosin II (heavy chain)	40 (rat), 5 (newt)	3 (Drosophila), 8 (nematode)		3 (Acanthamoeba), 0 (Dictyostelium)	
Glyceraldehyde phosphate dehydrogenase	11 (chicken)	1 (Drosophila), 2 (nematode)	7 (Aspergillus)	0 (Trypanosoma)	3 (maize)
Vittelogenin	33 (Xenopus), 33 (chicken)	1 (Drosophila), 0 (nematode)		0 (Dictyostelium)	
Actin	5–6 (rat), 5–7 (man), 6 (chicken)	0–1 (Drosophila), 1 (silkworm), 2–3 (sea urchin), 1 (nematode)	5 (Thermomyces), 1 (Saccharomyces), 0 (Schizosaccharomyces)	0 (Oxytricha), 0 (Dictyostelium), 0 (Tetrahymena), 0 (Entamoeba), 5 (Physarum), 1 (Acanthamoeba)	3 (soybean), 3 (maize), 3 (Arabidopsis)
Globin	2 (mouse)	2 (earthworm), 0 (Chironomous)			2 (soybean)
Alcohol dehydrogenase	9 (man), 9 (rat)	2 (Drosophila)	2 (Aspergillus), 0 (Saccharomyces)		9 (maize), 6 (Arabiodopsis)
Triose phosphate isomerase	6 (man), 6 (chicken)		5 (Aspergillus), 0 (Saccharomyces), 0 (Schizosaccharomyces)	0 (Trypanosoma)	8 (maize)
Phosphoglycerate kinase	10 (man)		1 (Aspergillus), 0 (Saccharomyces)	0 (Trypanosoma)	

base-pairs, the basic coding unit, and other collagen genes whether members of a family in the one organism or ones in different vertebrates follow the same pattern. Other genes that are thought to have been built up by such a method are those for immunoglobulins, α-fetoprotein, ovamucoid, growth hormone and ß-crystallin. The probable mechanism for the evolution of all these proteins having repeated structures is the amplification of a small ancestral gene by recombination within intervening sequences.

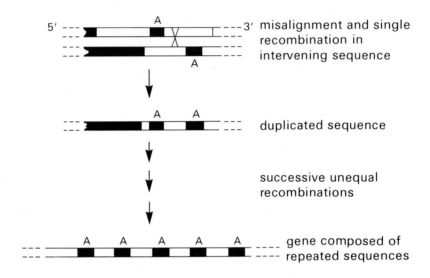

Figure 27 A model of gene amplification by recombination within intervening sequences. The letters are gene symbols.

A second proposed role in evolution for intervening sequences is the provision of sites for recombination between alleles. The major histo-incompatibility genes of man make membrane glycoproteins possessing three external protein domains or regions, a_1, a_2 and a_3, concerned with antigenic determination and separated by intervening sequences. Some human individuals have serologically different hybrid molecules composed of an a_1 domain coded from one allele, and the a_2 and a_3 domains from another. Nucleotide sequencing shows that the hybrid molecules are made from a new allele, probably produced by a single recombination between the intervening sequences of two different alleles (Holmes and Parham, 1985). From the number of silent substitutions that had occurred in the two alleles, the recombinational event was estimated to have occurred during the last 330 000 years.

A third evolutionary function of intervening sequences is the construction of new genes by the assembly of 'modules' of coding information contained in earlier evolved genes. Such a mechanism was predicted by Gilbert (1978) who considered that the presence of intervening sequences could increase the rate of protein evolution by a factor of 10^6 to 10^8. Evidence substantiating this conjecture comes from an examination of a human cell surface receptor gene (Südhof et al., 1985). The gene contains intervening sequences separating 18 coding regions, most of which specify functional domains. Thirteen of these regions encode sequences closely related to those in other known proteins; five encode a sequence homologous to that in the C9 component of complement, three code for a sequence present in epidermal growth factor and some blood-clotting proteins, and five specify sequences shared with the hormone for epidermal growth. Although the functional reasons for the piece-work assembly of this gene are not completely elucidated, it is evident that this receptor gene contains informational sequences shared with at least three other genes. Another possible example of this process is the genes for phenylalanine and tryosine hydroxylases in man and rat, which appear to have homologous regions coding for catalysis and nonhomologous ones for specificity (Ledley et al., 1985). Other genes suspected to have evolved in this fashion are those for glyceraldehyde phosphate dehydrogenase in fowls (Stone et al., 1985) and those for the proteases of blood coagulation and fibrinolysis in mammals (Patthy, 1985).

During messenger RNA processing, the intervening sequences are excised and the coding sequences are ligated in the same order as they were transcribed. An obvious alternative to assembling new genes is to make two or more proteins from the same template by ligating the RNA coding units in a different order (Figure 24, page 108). Such a device is commonly employed by some animal viruses and is also used in eukaryotic gene expression, but to an unknown extent. One example is the antibody genes of B-lymphocytes where the template for translation changes by switching alternative coding sequences. The gene first produces antibodies with tails to anchor them to cell surfaces where they act as receptors of foreign molecules. Then it changes to produce tailless antibodies which enter the blood stream, where they destroy the foreign antigens (Rogers and Wall, 1984). It does this by producing two different messenger RNAs, one giving a protein that is held in the membrane and the other a protein that is secreted. Another example is in the encoding of the mammalian hormone, calcitonin (Rosenfeld et al., 1983). This protein hormone is translated from one messenger RNA in the thyroid and a different one in the hypothalamus as a result of differential ligation of segments of the primary gene transcript. Consequently, the thyroid protein is a hormone active in calcium regulation, while that of the hypothalamus behaves like a neurotransmitter and is thought to be involved in cardiovascular and other control. While this particular strategy cannot be regarded as one which aids further evolution, it is a substitute for certain lines of tissue-specific gene evolution that would otherwise be necessary.

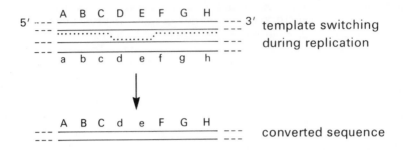

Figure 28 One of several possible models for gene conversion. The letters are gene symbols.

3.4.2 *Gene conversion*

This process has been shown to be a major source of genetic change in multigene families (Maeda and Smithies, 1986) which, according to the estimate of Hopkinson et al. (1976), is the state of about one-third of the enzyme-coding genes of man. Gene conversion consists in the replacement of a particular DNA sequence by another one (Figure 28), either between alleles at a locus, genes on the same chromosome, or genes on different chromosomes. As far as we know, the process requires sequence homology to operate but not gene contiguity, so the known examples are mainly in allelic genes or in families of closely related genes. After a gene is duplicated, the two nonallelic copies may diverge by nucleotide substitution, deletion and addition, but at a much slower rate than homologous genes from different species (Hood et al., 1975). This is because conversion between duplicate or multiple copies acts to maintain a uniformity of sequence. Conversions may be either very short, affecting only tens of nucleotides, or long, where several thousand nucleotides are changed.

Gene conversion can also spread an advantageous mutation to other members of the gene family. This has occurred in *Chlamydomonas*, as shown by the appearance of identical nucleotide substitutions conferring resistance to atrazine in the duplicated chloroplast genes coding for one of the membrane proteins (Erickson et al., 1984). Thus gene conversion provides a mechanism whereby a mutation may readily be duplicated or deleted without changing gene size or number. Gene conversion has also been detected in mammals (Slightom et al., 1980) and in yeast (Klein and Petes, 1981), where its frequency was estimated at 4% per generation for the loci involved. This frequency is much higher than that of spontaneous mutation, yet Thuriaux (1985) found that some yeast mutants have the frequency of intragenic conversion further increased by an order of magnitude. In copper-resistant strains of yeast, it was reported that a single gene for a copper binding protein was amplified by gene conversion to form ten tandemly arrayed copies (Fogel and Welch, 1984).

Gene conversion may acquire even greater evolutionary significance in conjunction with other processes. Thus, Levin (1983) and Reanney (1984) suggested that conversion may act upon the reverse DNA transcripts made by retroviruses which have incorporated cellular RNA, to integrate such sequences back into the host genome. Besides these general roles for gene conversion, the process seems to have been modified with respect to site specificity and frequency to control the mating-type switching that occurs in homothallic yeasts (Harber, 1983) and perhaps the pilus antigen gene switch in *Neisseria* (Meyer et al., 1984).

It is unlikely that gene conversion is the means whereby new genes are made by the assembly of parts of old genes because the apparent requirement for sequence homology restricts the process to alleles or duplicated genes. Nor is it likely to be a mechanism specifically evolved to aid evolution, but may rather be a byproduct of the processes of pairing and recombination between homologous chromosomes.

3.4.3 *Reverse transcription*

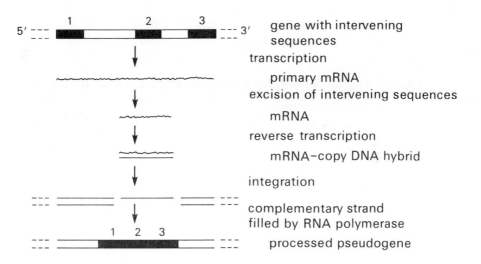

Figure 29 A model of reverse transcriptional integration. The numbers are coding sequences within a gene.

A third cause of genome reorganisation, demonstrated in yeast, insects and mammals, is the reverse transcription of RNA into DNA (Figure 29). In this process, the messenger RNA, after removal of the sequences corresponding to DNA intervening sequences and rejoining, is reverse transcribed from RNA to DNA, which may on occasion be integrated into the chromosome DNA. Its involvement in mammals is detected by the occurrence in the DNA of processed pseudogenes which have evidently been produced by the reverse transcription of messenger RNA. They are altered copies of functional genes elsewhere in the genome and resemble translatable messenger RNAs in lacking intervening sequences and in having 3'-polyadenine sequences, but they are usually not transcribed because they lack a transcriptional promoter (Hollis et al., 1982). These, plus other known or putative examples of reverse transcriptional addition to the genome (small nuclear DNA pseudogenes and most of the dispersed moderately repeated DNA of mammalian genomes), suggest that as much as 10% of the genomes of rodents and primates may have arisen by reverse transcription (Baltimore, 1985). However, it may be mainly in mammals that reverse transcription has generated pseudogenic sequences; they appear to be rare or absent in invertebrates (Rogers, 1985) and it is not yet clear whether they exist in nonmammalian vertebrates (Table 13).

TABLE 13
The distribution of the processed pseudogenic forms of several genes (data from Wagner, 1986). Cellular reverse transcriptase has not been found in any organism, so retroviruses may be required in pseudogene formation. Retroviruses occur in mammals and birds, but only one pseudogene has so far been found in birds.

	Mammal	Vertebrate Bird	Amphibian	Invertebrate (Drosophila)	Fungus (Saccharomyces)
Ribosomal proteins	+		−	−	−
Actin	+	−		−	−
Tubulin	+	−	.	−	−
Cytochrome	+	−	−	−	−
Glyceraldehyde 3-phosphate dehydrogenase	+	−		−	

As with conversion, the evolutionary role of reverse transcription may lie in its interaction with other processes of sequence rearrangement. For example, Lueders et al. (1982) have noted that a mouse a-globin pseudogene lacking intervening sequences has retroviral-like sequences adjacent to it in the mouse genome, suggesting that the globin gene was transposed via spliced retroviral RNA. It has been suggested by Reanney (1984) that new genes could be built up, following removal of the intervening sequences in messenger RNA, by novel conjunctions of coding units and reintegration in DNA by reverse transcription, or by gene conversion. Such a mechanism for the construction of new genes would be alternative, or additional, to their assembly by recombination at intervening sequences.

3.4.4 *Transposition*

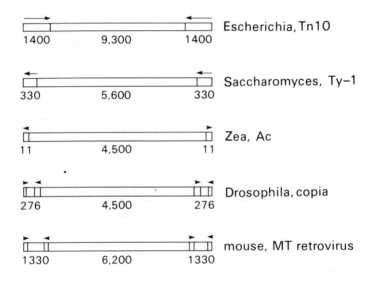

Figure 30 Diagrams of the structure of some prokaryotic and eukaryotic transposable elements. The arrows show the direction of the repeated sequences and the numbers are the sizes of the segments in nucleotide pairs. Tn (Foster et al., 1981), Ty-1 (Farabaugh and Fink, 1980) copia (Levis et al., 1980), mouse mammary tumour virus (Varmus, 1983).

A final known agent which is frequently thought to be significant in eukaryotic evolution is the various transposable elements (transposons) and possibly related retroviruses (RNA viruses, DNA copies of which can integrate into cellular DNA). The eukaryotic transposons and integrated retroviruses (proviruses) are sequences of DNA a few thousand base-pairs long, possessing sequence repeats of varying length at each end which appear to be regions involved in transposition (Figure 30). Their evolutionary role may be enhanced by the fact that at least some of them are either transmitted through the germ-line (retroviruses) or are preferentially active there (*Drosophila* transposons). Indeed, Kimura (1985) has remarked that, while the accumulation of nucleotide replacements is constant per year, the generation-dependent occurrence of lethal and 'visible' changes is now suspected to be primarily caused by transposable elements. Certainly, a large proportion of spontaneous mutations is caused by the insertion or deletion of movable sequences in yeast (Roeder and Fink, 1983), *Drosophila* (Rubin, 1983) and maize (Wessler and Varagona, 1985). The high frequency of these events (accounting for 25 to 65% of mutations in the genes examined) reflects the relatively large amount of movable sequence in the genome; about 10% in yeast (Rubin, 1983), 5 to 10% in *Drosophila melanogaster* (Finnegan, 1981) and about 0.1% of endogenous proviruses in rodents and primates (Todaro et al., 1980). However, the actual amount of movable sequence in a species does not seem to be important. For example, *Drosophila melanogaster* has about 20 million base-pairs of mobile genetic elements or their defective relatives, organised into 30 to 50 families. In contrast, its sibling species, *Drosophila simulans*, which is almost identical morphologically to *D. melanogaster*, has only about three million base-pairs that are mobile (Dowsett and Young, 1982).

Some movable elements of the 'retroposon' class may do more than merely inactivate other genes when they transpose. Presumably, because of promoters, terminators etc. in their long terminal repeated sequences, they can have various effects on transcription, such as enhancing or depressing it or polyadenylating the transcripts (Section 2.2.12). In yeast, the transposition of the element Ty-1 to the 5' region of the alcohol dehydrogenase gene changes its expression from being glucose-repressed to constitutive (Williamson et al., 1981). Transposition probably does not cause base substitutions, but recombination between transposed elements in different locations can generate inversions, deletions and duplications. The activation that occurs in certain *Drosophila* hybrids, and which induces chromosomal and developmental abnormalities, could initiate reproductive isolation and thus species formation (Rose and Doolittle, 1983). The significance of the eukaryotic movable elements may lie in promoting the evolution of the species and not the survival of the individual for, as far as is known, they do not seem to confer selective advantages on the hosts carrying them.

Evidently the eukaryotes contain a number of systems for the rearrangement of DNA sequences, but we do not know whether any of them act in a nonrandom manner. The most significant for the production of new genes appears to be the rearrangement of gene parts that may depend on the presence of intervening sequences. But a mechanism giving similar results is detectable in bacteria, where intervening sequences cannot be involved in the process because they are not present in bacterial protein-coding genes. Moreover, it is not clear how a rearrangement of gene parts can occur in the absence of sequence homology.

3.5 *Conclusion*

As far as can be deduced, both the origin of life and much of its early evolution are attributable to selection acting on variations produced by the random and damaging mechanism of point mutation. However, the apparent evolutionary ineffectiveness of mutation in existing organisms indicates that a time came when the possibilities for further significant evolution by mutation became exhausted, or at least very limited. Systems that evolved to reduce the frequency of spontaneous mutation and to minimise its damaging effects, which would become more pronounced as the organisms became more complex, also reduced its efficacy. Consequently, evolution by mutation must have slowed down, generating high selection pressure for organisms which could evolve faster by means other than the gradual accumulation of advantageous mutations. The development of new mechanisms of evolutionary change would become possible only when a reservoir of informational nucleic acid sequences had built up over tens or hundreds of millions of years by the processes of random mutation and selection.

It is likely to have been against such a background that additional means of generating adaptive variation evolved, means which increased both the frequency and nonrandomness of genetic change, perhaps with a lessened deleterious component. The devices then employed can be seen today in a highly evolved form in the substructure of genes and in the presence of mechanisms for gene movement, assembly and transfer. This is not to say that evolution by nucleotide substitution has been discarded; it is still a significant component of the evolutionary process, one concerned with the fine adjustment of proteins and other gene products to special features of the cell or organism. And it should be remembered that what we have found is an inability of mutation to generate new substrate specifications in enzyme molecules; we do not know how effective it may be in affecting interactions between proteins and between proteins and nucleic acids.

The various nonmutational mechanisms that have been mentioned (horizontal transfer, transposition, conversion, reverse transcription, modular assembly) can together provide an extensive amount of genetic variation. The average frequency of transposition alone is about 5×10^{-3} per transposon per cell in bacteria (Kopecko, 1980), about 10^{-4} per generation for transposable elements in *Drosophila* (Ising and Block, 1981) and from 10^{-4} to 10^{-8} per element per gamete in maize (Pryor, unpublished). Even if these sequence alterations were no more advantageous to the organism than point mutation and chromosomal rearrangement, their combined frequency of occurrence is much greater. Moreover, whereas mutation causes quite random changes in genes, many of the other mechanisms introduce or rearrange ordered sequences of genetic information. Evolution by the rearrangement of earlier evolved genes or gene parts may give rise to quite large changes in phenotype which are viable, unlike the results of point and chromosomal mutation where changes of major effect are usually deleterious, at least in animals.

121

There is not yet sufficient information to say whether or not new evolutionary mechanisms have come into play with phylogenetic advance. The most that can be said is that different groups of organisms have adopted or emphasised somewhat different evolutionary strategies during the course of their evolution. Apart from the unsegmented RNA viruses which may still rely mainly on mutation, the other viruses and the prokaryotes use various forms of horizontal gene or subgenic transfer for their evolution. The means adopted change from DNA bacteriophages, where whole blocks of genes are exchanged by recombination between members of related families, to the more advanced bacterial systems in which recombination is not required because of the autonomous replication of plasmids. In the eukaryotes, horizontal gene transfer is uncommon and intragenomic transfer of subgenic segments by intervening sequence recombination and perhaps gene conversion and reverse transcription may be one of the main mechanisms, although the details are not yet clear.

PART 4
The Evolutionary Potential of Living Organisms

A question of comparative interest is: are the evolutionary changes described in Part 2 occurring in all organisms or only in certain groups? The answer depends on whether, as has been claimed, there has been a reduction in the diversity of evolving life forms or a general decline in the processes of evolution. In this part of the book, we will attempt to find an answer by making very approximate assessments of the potential for novel evolutionary change in the major phyla of organisms.

4.1 The Continuity of Evolution

A unique property of all organisms is their ability to become less random, or more organised, with time (Blum, 1955). This is another way of saying that organisms have the capacity to evolve, and it is one of the criteria by which living matter can be distinguished from the nonliving. Evolution is not a process which has occurred only in the past; it is still going on today and should continue for many millions of years into the future. Such is the generally held view. In fact, Haldane (1954), after considering the violent climatic changes of the Pleistocene and the influence of man, suggested that evolution may now be proceeding with unparalleled speed.

There is, however, information which may indicate that evolution is not now the progressive process that it once was. It is apparent that the origin of the so-called 'urkingdoms' of organisms, Eubacteria, Archaebacteria and Eukaryotes, finished at least 1400 million years ago when the eukaryotes appear in the fossil record. All the major prokaryotic groups were already present in the Proterozoic, and those of the Protista in the Carboniferous. Animal evolution on the larger scale, phyletic evolution, apparently ceased with the origin of the Agnatha (jawless fishes) about 500 million years ago. No new class of protist, plant or animal is known to have appeared since the Jurassic (150 million years ago) when the reptile-bird transition occurred (Figure 31).

These considerations give the impression that evolution is running down:

that the evolutionary force is steadily weakening or that the evolutionary potential of life is becoming used up. Evolution may be already finished in the grander sense and may now be confined to the elaboration and diversification of a limited number of adaptive strategies which evolved a long time ago.

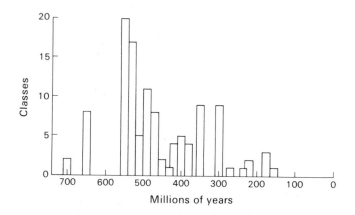

Figure 31 The time of origin of the classes of eukaryotes, including vertebrates, invertebrates, plants and protista but excluding algal classes and the subclass Copepoda of the class Crustacea, which have poor fossil records.

The fossil evidence is not alone in suggesting a decline in the achievements of evolution. The mechanisms of evolution are directed, generally speaking, to the adaptation of an organism to components of its environment. In the process, the structure, physiology and behaviour of the organism come to fit ever more closely the peculiar features of its particular environment. This is a matter of specialisation whose concomitant is usually a decrease in adaptive flexibility because the organism has given up its ability to live over a range of environments. Also it may have become lethargic or even sessile, with well-developed external skeletons or other special features of passive defence. Where evolution is still continuing, that is, where natural selection is intensive, most evolving groups comprise wide-ranging organisms which prey upon lesser ones and are preyed upon by others. But when evolutionary equilibrium or stasis is reached, it results in a very closely adapted organism in which, however, the potentiality for further change is markedly decreased, frequently by the confines of specialisation.

A further cause for the slowing of evolution is undoubtedly the filling of most of the major ecological zones. Rensch (1959) compared similar habitats in different regions of the world with respect to the frequencies

124

of forest birds of different sizes, terrestrial snails of different families, successions in giant animals and carnivorous mammals, and soil organisms of various phyla. From the fairly constant percentages of different animal types in a variety of geographical regions and geological epochs, he concluded that most habitats have long been saturated with life forms. Terrestrial habitats tend to be somewhat less saturated than aquatic ones and it is on land that the process of evolution appears to be generally more rapid. While evolution was initially directed to adapting life to new unoccupied habitats, as in the development of land plants, it later served primarily as a means of replacing one type of organism by another, as with the fishes. Then, as potential hosts became diverse and abundant, it facilitated commensalism and parasitism, frequently by retrograde evolution, as shown by the protozoan and worm phyla. As available environments for new life forms have become more restricted, so has the scope and rate of evolution lessened. Indeed, quoting the numerous occurrences of stationary, relict or parasitic species, Decugis (1941) concluded that life manifests many signs of the onset of senescence.

4.2 *Evolutionary Stasis*

It is possible, at least in theory, that the overall course of evolution is one in which the earlier and simpler organism remains in equilibrium after having given rise to more complex forms. As time goes on, more and more species reach equilibrium until all environmental niches are occupied and evolution then stops. Mainly for this reason, Huxley (1943) saw evolution as a series of adaptive experiments of greatly varying duration, all of which, however, terminate blindly. An opposing view, also theoretical, is that evolution, or more precisely genotypic adjustment, can never stop because a perfect equilibrium between life and its environment cannot be reached. This argument rests on the proposition that a species is so highly adapted to its environment that any occurrence of spontaneous mutation, change in climate or increase in the competitiveness of other species will be deleterious (Fisher, 1930). Hence the environment of a species is continually deteriorating and the species must make repeated genetic adjustments to compensate for the deterioration.

There are, however, certain mechanisms whereby an organism may adapt to a changed environment without undergoing genetic change. Homeostasis or buffering acts at the macromolecular, physiological and developmental levels to ensure that the phenotype predetermined by the genes is adhered to in the face of environmental exigencies tending to distort it. The obverse is phenotypic flexibility, a device of the genotype to allow short- or long-term shifts in phenotype in response to environmental change. An additional strategy employed by some organisms is to pre-evolve a mechanism which adequately copes with fluctuations of an important component of the environment. Examples are phase variation in *Salmonella* and other bacteria, mating-type switches in yeast, coat protein changes in trypanosomes and antibody diversity in mammals. In these instances, environmental change does not result in a change in the genotype, which is biologically expensive, but merely in its qualitative expression.

4.3 *Criteria in Assessing Evolutionary Potential*

The evolutionary possibilities for a group of organisms may be estimated from the scope of its environment, whether narrower or broader, the specialisations involved in adapting to it, whether inhibiting or enhancing further progress, the general trends exhibited by its past evolution, including the nature of adaptive changes in recent times, and finally, where possible, the status of its evolutionary mechanisms. For example, with regard to environmental range, whereas the Dipnoi (lungfish) are confined to small wet regions subject to seasonal drought in Africa, Australia and South America, the amphibians are widespread but dependent on water for reproduction, and the rodents are abundant the world over. The lungfish have undergone very little change since the Triassic, the more advanced Amphibia, the frogs, have remained much the same since the Cretaceous, while the rodents, which originated at the end of the Palaeocene, seem still to be evolving. Both the lungfish, represented today by three genera, and the amphibians, with three living orders, are groups which have steadily decreased in number and variety with numerous extinctions, but the rodents are among the most successful of mammals and have suffered relatively few extinctions. These three criteria are consistent in placing the evolutionary potential of these groups in increasing order, as Dipnoi, Amphibia, Rodentia.

There remains a further and important criterion to consider, that of the degree of specialisation already attained. By specialisation, which occurs as a normal part of evolutionary progress, is meant adaptation for a specific function or environment. It has frequently been remarked, e.g. by Carrington (1956), that periods of rapid evolutionary radiation from simple readily adaptable forms are regularly followed by relatively long phases of increasing specialisation. A particular specialisation may be morphological as in the limbs (wings and claws) of birds, ecological as in the red algae with their narrow range of marine depths, biochemical as in pressure-adapted enzymes of deep-sea fishes, or genetical as in the apomictic reproduction of many plants. With respect to evolutionary potential, the palaeontological evidence demonstrates that the development of new large-scale innovations in function or structure usually come from lines that are still relatively unspecialised (Huxley, 1955). Hence the 'law of the unspecialised' of Cope (1896), which states that evolutionary advance is possible only to simple unspecialised forms and that specialisation causes evolution to slow down and eventually stop; extinction commonly follows. This view has been supported by Rensch (1959) who noted that the most frequent causes of extinction are superior trans-specific competition and highly increased specialisation, the two often being connected.

All organisms can be said to be specialised to some degree, but while the specialised features may lack evolutionary potential, this need not apply to the organisms themselves. Thus in marsupials, both brush-tailed possums (*Trichosurus*) and koalas (*Phascolarctos*) are thoroughly specialised for an arboreal existence. While the possum can live in treeless areas, eat a wide variety of plants and breed once a year, the koala is sedentary, eats leaves of a very few species, has a peculiarly specialised digestive system and is slow-breeding. The evolutionary potential of the possum is still high, as is also indicated by its three species and seven subspecies, whereas that of the koala, a monotypic genus, is very low because of its extreme specialisation. It is not usually a single specialisation, or even a few, that block further evolutionary advance, but a whole complex of specialisations.

4.4 *Comparative Aspects*

The major evolutionary advances such as those that typify a phylum, class or order involve the acquisition of evolutionary novelties, qualitative changes in a structure or property, which ensure an additional function of an organism, enabling it to adapt to new environments. The taxon taken to be indicative of a novelty, a new and distinct adaptation, is the order. Once an order has arisen, its novel design is usually distributed into lower taxonomic categories by processes of modification and elaboration.

If the origin of orders is followed from the time of origin of the phylum or class to the present day, comparisons may be made between periods within lineages to provide a rough guide to the overall course of evolution. Some investigators have advocated a comparison of periods of equal duration, presumably for the purpose of showing that rates of novel evolution differ between lineages, as they certainly do. However, it is comparisons within lineages, not between them, that are relevant to the question of evolutionary potential. Nevertheless, it is important to know if the taxa have an equivalent status in different phyla.

It is generally agreed that phyla are anatomically comparable, although there may be too many small phyla of invertebrate animals (Van Valen, 1973), because workers at the level of phylum must make comparisons with other phyla. Similarly, it seems likely that species are defined comparably for most groups of sexually reproducing organisms, except perhaps for the ciliates and for a tendency to regard some subspecies of angiosperms as distinct species. Most differences are in the taxonomic divisions between phyla and species, with a divergence in comparability usually being greatest for the lower taxa. For example, DNA hybridisation data suggest that bacterial families are more distinct than vertebrate classes (Hoyer et al., 1964), on the basis of serological comparisons, frog genera are as diverse as families or suborders of placental mammals (Wallace et al., 1971) and anatomically there is no greater difference between orders of birds than exist between families of other vertebrate classes (Romer, 1966).

Kurtén (1969) has proposed that in each of the major land masses, about six orders of mammals and reptiles have evolved, implying that the orders are similar in adaptive divergence. The tendency in animals is to divide orders into families, but in angiosperms the procedure is to group families into orders, and for this reason these taxa may not be equivalent in the two kingdoms.

Evolution is usually said to 'progress' or 'advance' and organisms are referred to as 'lower' or 'higher' and by similar comparatives, although these terms cannot be logically sustained. To speak of evolutionary 'advance' gives a teleological flavour for it supposes that earlier stages are intermediates in the production of later ones in a kind of progression to a predetermined end. The corollary, subject to the same evaluative implications, is that earlier evolved organisms are simpler or more primitive than later evolved ones. Nevertheless, the use of such terms only in a historical sense, and with a recognition of their arbitrary nature, can be justified as a matter of convenience.

It is sometimes proposed that, although there may be little morphological change, active metabolic evolution may be occurring in arrested groups. This seems unlikely, because the main elements of metabolism, together with their adjusting systems, were fully evolved long ago. On the other hand, it is probably not valid to assert that evolution in the Phanerozoic (the last 600 million years or so) has been proceeding at a much more rapid rate than in the preceding Proterozoic. The bias of the observer is towards change in morphology, so less account is taken of the equally significant evolution that provided the intricate details of genetic organisation and transmission, metabolic pathways and cellular structure.

4.5 *Evolutionary Assessment of the Major Phyla and Classes*

4.5.1 *Prokaryotes*

The main groups of prokaryotes seem to have been fully evolved before the start of the Cambrian period (Walter, 1987). Anaerobic photoautotrophic bacteria were present 3500 million years ago and sulphate-reducing bacteria, cyanobacteria and archaebacteria appear in the Archaean eon; the Actinomycetales (branched filamentous bacteria) are found later in the Proterozoic. All these prokaryotic groups are still extant and appear to be much the same now as they were then.

In the early prokaryotes, the major trend in evolution was probably towards the development of diverse energy transformation systems using the mechanisms of mutation and perhaps recombination which are little employed by present prokaryotes. This evolutionary epoch provided most of the metabolic systems needed for higher forms of life to evolve. Subsequent evolution has taken the direction of close and efficient control of these energetic and synthetic processes, made necessary by intense selection for poorly and sporadically available energy sources. Substrate limitations also provide selection pressure for speed of reproduction, reaching rates of 8.5 minutes per generation in some thermophiles. To aid rapid reproduction, the coupled system of transcription and translation was evolved, or retained, thus preventing the possible use in evolution of modular assembly of genes based on the presence of intervening sequences (Gilbert, 1978).

Still later, certain systems were evolved, based on past reaction to environmental change, which provided adequate responses to many environmental fluctuations that were likely to occur. Such systems comprise transposons and genes carried on plasmids, internal insertion sequence and transposon movement, and phase and pilus variation by template switching. In addition, the prokaryotes have developed various nonspecific mechanisms which resist the requirement for genetic change to environmental perturbations, e.g. strong cell walls, homeostatic metabolic mechanisms and spore formation in adversity. These are all devices for reducing the necessity for formal evolution, e.g. by mutation and selection.

Natural populations of enteric bacteria contain larger amounts of genetic variation in the form of multiple alleles at many loci than do populations of eukaryotes (Selander and Levin, 1980). Although mechanisms comparable to sexual recombination, namely conjugation, transduction and transformation, exist in bacteria, the large number of genotypes comprising a population seldom recombine. The population structure is essentially one of numerous independent lineages or clones (Ochman and Selander, 1984). Each clone is characterised by having one or more mutations substituting amino acids in enzymatic proteins; they usually appear to be selectively neutral (Hartl and Dykhuisen, 1984), and consequently have limited evolutionary significance.

Evolution in the prokaryotes today seems to be much more limited in one sense, that of producing new genes, than it must have been in the past. The sense in which it is now more efficient is in the acquisition of the genes of other prokaryotes. An indication of the efficiency of plasmid transfer in response to the prevalence of antibiotics is given by Bulling et al. (1973). Over a ten-year period, antibiotic-resistant strains of *Salmonella* from animal sources rose from 2.5% to 9.5%. There was also an increase in the spectrum of resistance; doubly resistant strains rose from 8% to 25%; 8.5% had triple resistance and 1.5% had quadruple resistance; no strains had resistance to more than two antibiotics initially. The evolutionary strategy adopted by the bacteria is that in which nearly all bacterial genomes form part of an extended pool of genetic information which a bacterium can draw upon as the need arises. It is a very powerful strategy allowing for rapid adaptive change, but limited in scope because of its reliance on the products of earlier evolution.

The archaebacteria, eubacteria and at least the nuclear component of the eukaryotes all appear to have originated from a common ancestor, perhaps 3500 million years ago (Walter, 1987), at a time when the hydrosphere is thought to have become oxidising (Walker et al., 1983). The genetic nature of the common ancestor or progenote can in part be deduced as having those features that are found in prokaryotes and the nuclear part of eukaryotes, and that are therefore present in the progenote. Although information is still sparse, this line of reasoning may suggest how some genes first came in being. The enzymes of energy metabolism such as triose phosphate isomerase, phosphoglycerate kinase, alcohol dehydrogenase and glyceraldehyde-3-phosphate dehydrogenase, as well as the serine proteases, are necessary in even the most primitive organisms for essential energy transformations and for the metabolism of their own proteins.

For example, the proteases cleave peptide bonds by a polarisation of the bond by a nucleophilic amino acid assisted by a proton-donating amino acid at the active site. The three catalytic amino acids, histidine, serine and aspartic acid, and the primary and secondary binding sites, are coded for by five of the eight sequences separated by intervening sequences in the eukaryotic protease (Campbell and Porter, 1983). The structure is consistent with an origin for the primordial protease gene by the joining together of nucleic acid sequences coding for protein domains of perhaps 15 to 20 amino acids which have become expanded by the sliding of intervening sequences and by insertions to give domains of 33 to 36 amino acids in contemporary proteases. Subsequently, duplication, deletion, insertion and mutation of either the entire gene or of its protein-coding domains have altered the primordial protease gene to make other digestive proteolytic enzymes, enzymes of the blood clotting and complement cascades, protease inhibitors and a number of other polypeptide-processing enzymes involved in pain sensing and fertilisation. This example of primitive protease structure indicates that intervening sequences were present in at least some genes of the ancestor of prokaryotes and eukaryotes and that recombination occurred within them. The same conclusion was reached by Quigley et al. (1988) from comparison of the position of intervening sequences in the gene for glyceraldehyde-3-phosphate dehydrogenase.

Other gene products common to prokaryotes and eukaryotes, but of which less is known, include the DNA-binding proteins or polypeptides in prokaryotic repressors and in the hormone receptors of vertebrates (Baker, 1988). The transcriptional RNA polymerase of bacteria and the three eukaryotic polymerases share a common subunit structure in the basic enzyme (Armaleo, 1987), and a protein assisting protein folding in bacteria and archaebacteria is homologous to one in *Tetrahymena* which probably assists the refolding of proteins newly imported into mitochondria and one in plants that helps the assembly in chloroplasts of subunits of ribulose bisphosphate carboxylase (Hemmingsen et al., 1988).

With regard to the evolution of metabolic pathways, most of those for energy transformations and the biosynthesis of essential amino acids, carbohydrates, vitamins and nucleotides would be expected to be present in the prokaryote-eukaryote precursor. Seven of the eukaryotic enzymes in the glycolytic pathway have a catalytic domain separated by an intervening sequence from a coenzyme domain to which adenosine triphosphate or nicotine adenine dinucleotide binds. The nucleotide-binding site in several of the enzymes has a very similar conformational (three-dimensional) structure but shows little sequence homology, while the catalytic domains are unique for most of the enzymes. The structure of this nucleotide-binding sequence suggests that the gene was built up by successive duplications of a simple sequence encoding an α ß coding unit to give a three-fold repeat capable of encoding a mononucleotide subdomain. Further recombination within flanking intervening sequences would duplicate the three-fold cluster to produce an ancestral dinucleotide binding domain (Michelson et al., 1985) which combines with individual catalytic domains by gene fusion. This conclusion is derived from the similarity of coding sequence patterns in mammalian glycolytic genes and the correlation of intervening sequence locations with domain and subdomain boundaries. A similar distribution of substrate-binding and catalytic domains, but without the intervening sequences, is found in some glycolytic dehydrogenases of *Escherichia coli,* suggesting that very early cells before the separation of prokaryotes and eukaryotes probably had split genes containing intervening sequences.

An indication of the molecular mechanisms that are presumably still available for evolution may be obtained by considering genetic means of adaptation after the origin of prokaryotes. The examples are mostly of interactions between eubacteria and later arising eukaryotes.

Some subspecies of *Bacillus thuringiensis* produce a δ-endotoxin during sporulation that is extremely toxic to the larvae of blackflies and mosquitos. The responsible gene may have been acquired by means of a transposon since the endotoxin gene in *Bacillus thuringiensis* is flanked by inverted repeat sequences (Kronstadt and Whiteley, 1984). Horizontal gene transfer by means of plasmids is also indicated for the acquisition by enterotoxigenic strains of *Escherichia coli* of plasmid-borne genes for intestinal toxins from the cholera-producing bacterium, *Vibrio cholerae.* The number of synonomous (silent) nucleotide substitutions in the genes from the two bacteria show that they diverged about 130 million years ago when *E. coli* acquired the toxin gene, whereas *E. coli* and *V. cholerae* diverged about 670 million years ago (from 5S ribosomal RNA data, Yamamoto et al., 1987).

The spirochaete eubacterium, *Borrelia hermsii,* causes relapsing fever in man and analogous diseases in other mammals, reflecting phases of antibody formation against the bacterial surface proteins which become temporarily ineffectual as the bacterium changes its antigenic amino acid sequence. The rate of switching between serotypes is estimated to be 10^{-4} to 10^{-3} per cell per generation, resembling the antigenic variation of the eukaryotic protozoan, *Trypanosoma*. The mechanism of DNA rearrangement in *Borrelia* appears to be by the duplicative transposition of a storage copy of part or all of an antigen-specifying gene to an expression site (Meier et al., 1985). The transfer of information is probably by the process of gene conversion, as has been shown to be the mechanism for antigenic variation in *Neisseria gonorrhoeae* (Haas and Meyer, 1986). Antibodies to this bacterium are usually directed against pilin, the protein subunit of the pilus, and against the opacity protein of the outer membrane, both of which are involved in adhesion of the pathogen to the human mucosa. However, the antigen-antibody reaction may be overcome by phase variation when the antigenic sequences spontaneously alter with high frequency. Both of these modes of variation are associated with DNA rearrangements in two expression loci of the pilus structure, the remaining five pilus loci of this strain being incomplete and silent. The transfer of genetic information from silent loci to expression loci by gene conversion may be made unidirectional and greatly increased in frequency by the presence of short repetitive elements acting like recombination enhancers. In respect to conversion, the mechanism of antigenic variation in *Neisseria* resembles that in the surface proteins of *Borrelia* and in eukaryotes such as in the yeast mating-type system and the system for the variation of surface glycoproteins in *Trypanosoma*.

131

Conversion was probably also involved in making genes for enzymes carrying out sequential but catalytically diverse reactions in the ß-ketoadipate pathway in *Acinetobacter* and *Pseudomonas*, where relatively short regions of sequence homology have been transferred from one set of structural genes to another (Yeh and Ornston, 1980). A similar example is in the assembly of some *Bacillus* spore proteins by rearrangement of blocks of primary amino acid sequence (see Section 3.3.2). Since there was no change in gene length, the DNA sequences were inserted by substitution, not addition, which apparently occurred without the usual requirement for sequence homology, but perhaps with the help of recombination enhancers. These evolutionary events may have occurred before the eukaryotes arose, but there is evidence that the mechanisms have operated in the recent past. Strains of *Neisseria gonorrhoeae* that do not produce a penicillin-destroying ß-lactamase acquire resistance by making a penicillin-binding protein with little affinity for the antibiotic. The penicillin-binding protein is an enzyme concerned with peptidoglycan synthesis for the cell wall, and consists of a penicillin-insensitive transglycosylase domain and a penicillin-sensitive transpeptidase domain. Reduced affinity in resistant strains is achieved either by the accumulation of point mutations that discriminate between the binding of antibiotic and natural substrate or by the replacement of part of the transpeptidase domain of the enzyme with a mutated low penicillin-affinity version (Spratt, 1988).

A variant of such recombination-associated determination of gene structure and expression is the inversion of DNA segments by site-specific recombination (Craig, 1985). Such inversions serve different functions in prokaryotes, such as change in the host range of bacteriophage *Mu,* P 1 and P 7, variation in the flagellar antigen of *Salmonella* and inversion of unknown role in *Escherichia*. All these inversion reactions are closely related, as the element-encoded recombinases are highly homologous and the recombination sites have homologous sequences. In *Salmonella typhimurium,* for example, phase variation is brought about by the alternate expression of two nonallelic genes for flagellin, the protein subunit of the flagellum. Which gene is expressed depends upon the orientation of an invertible DNA segment that codes for the enzyme catalysing the inversion and is bounded by two short inverted repeats. The invertible segment also contains a promoter that, depending upon orientation, switches on one of the two flagellin genes outside the segment. These site-specific invertases are related in structure and sequence to the 'resolvase' of transposon 3, suggesting that the corresponding genes had a common ancestry. This relationship implicates transposon movement as being responsible for the insertion of site-specific recombination systems into different genetic structures as a type of module.

Some transposons themselves may contain recombinational hot spots or enhancers for the insertion of additional resistance genes. The ß-lactamase transposons of Gram-negative bacteria, in particular, often carry other resistance genes, such as one for streptomycin and spectinomycin resistance or for sulphonamide resistance. These recombinational hot spots have been sequenced and shown to be homologous to the consensus

132

recognition site of bacterial invertases, suggesting that a transposon-encoded enzyme may interact with the target site (Oullette et al., 1987).

The above examples demonstrate that, at the molecular level, many evolutionary events in bacteria are brought about by site-specific recombination within already evolved genetic systems. Thus recombination between inverted repeats results in the inversion of DNA segments as in host range and flagella variation. Recombination between repeat sequences on two DNA molecules causes co-integration of the molecules as in transposon insertion, whereas recombination between direct repeats on one DNA molecule brings about the excision of a DNA segment as in bacteriophage and transposon deletion.

Possible steps in gene evolution may involve, firstly, the *de novo* origin by nucleotide substitution of short informational nucleotide sequences, perhaps interrupted by noninformational stretches or intervening sequences; secondly, the multiple duplication of these primordial genes to give larger ones capable of coding for additional or intensified functions; and thirdly, the development of recombination, perhaps as a nucleic acid repair mechanism directed against damage induced by ultraviolet light. This step could enable the primitive genes of progenotes or proto-organisms to acquire pieces of sequence information from other genes via recombination in intervening sequences, to form the basic structure of early prokaryotic genes. Fourthly, as an aid to rapid cellular multiplication, intervening sequences were lost, thus bringing prokaryotic evolution to a virtual stop except for the acquisition of the genes of other bacteria carried on plasmids and transposons or as part of a viral genome. With a restoration or retention of intervening sequences in the genes of eukaryotes, new genes and new avenues of evolution may have became possible by subgenic recombination.

4.5.2 *Protista*

The protista are all lower eukaryotes which have levels of organisation above those of the prokaryotes but below those of vascular plants and animals. Despite their primitive differentiation and morphogenesis, the Protista lead a highly successful form of life, possibly because they occupy ecological sites unavailable to the larger higher eukaryotes.

The simpler and earlier eukaryotes are microorganisms classified either as protista or fungi. Members of the kingdom Protista are not fungi which develop from spores, nor plants which develop from an embryo, nor animals which develop from a blastula. They are very heterogenous organisms containing some 34 major phyla, according to Margulis and Sagan (1986). Very little is known of their biology, especially at a molecular level although ribosomal RNA sequences indicate that most of them arose during a relatively early radiative period in eukaryotic evolution. Sogin et al. (1986) suggest that this radiation may have been the result of new methods of organising or processing genetic information.

The probable order of origin of the main groups ('phyla') of the Protista, according to 16S ribosomal RNA sequences is: microsporidia, flagellated protozoa, euglenoid ciliates, cellular slime moulds, plasmodial slime moulds

133

and dinoflagellates. All these groups first arose in the early to mid Proterozoic eras before the appearance of the multicellular kingdoms, fungi, plants and animals. Some other groups of the Protista, like the ciliates, chrysophytes (golden-brown algae), oomycetes, green algae and pseudopodial protozoa, seem to have appeared after the differentiation of the three main kingdoms, probably in the late Proterozoic (Gunderson et al., 1987). The evolutionary succession deduced from 5S ribosomal RNA sequences is rather different (Hori and Osawa, 1987), but 5S sequences change at about twice the rate of 16S sequences and are unreliable for times greater than about 1000 million years (Ochman and Wilson, 1987).

Algae
The algae seem to be the oldest fossil group with ribbon-shaped megascopic forms of unknown affinities, being present 1300 million years ago (Walter, 1987), and with the green, brown and red algae being already differentiated by the Proterozoic (Loeblich, 1974). However, their fossil record is misleading, for almost all the fossil algae are calcareous, whereas most living algae are not. They constitute probably the greatest bulk of protistan living matter, since marine (algal) chlorophyll comprises 7.4% of the world's total (Lieth and Whittaker, 1975). They are also the least specialised and the most versatile of the Protista. From a relatively unspecialised ancestral dinoflagellate, of the form of *Gymnodinium*, have evolved various single modes of nutrition (photosynthetic, phagotrophic, parasitic or osmophilic), cell forms (walled, armoured or amoeboid) and organism structure (unicellular, coenocytic or partly multicellular) (Stanier et al., 1972). While many of these adaptive lines were probably terminally evolved, the ancestral dinoflagellate continues to exist and could presumably re-evolve.

In plant evolution, in addition to the retention of photosynthetic mechanisms, numerous lines of developmental organisation have arisen from the weak initial inventions of the algae. They include not only cellulose walls, cell polarity and multicellularity, but cell differentiation, shape depending on the plane of cell division and both apical and intercalary meristems. One line of evolution in multicellularity and differentiation reaches its speak in *Volvox*, individuals of which contain several thousand cells. However, this green algal line of advance failed to evolve further, possibly because the cells of the colony remained surrounded by gelatinous sheaths, thus preventing developmental mechanisms based on interactions between cells (Kochert, 1973). Other contemporary algae of the *Volvox* group are simple, apparently unspecialised organisms of the *Chlamydomonas* type, which are thought to have given rise to land plants and presumably have the capacity to do so again. As far as can be deduced, therefore, the evolutionary potential of the algae, in particular the green algae, is still high.

Protozoa
Some classes or orders of Protozoa (Foraminifera, Ciliata, Radiolaria) have been attributed to the Cambrian or Ordovician periods, but generally

their fossil record is poor and their taxonomy uncertain. The progenitors cannot now be identified, primarily because of the rapid and diverse evolution of the Protozoa along two main lines. One evolutionary line, which began in the Cambrian or earlier, has been towards the pseudopodial forms, such as amoebae, radiolarians and heliozoans. As a defence against other predators, some of these Protozoa have developed siliceous or calcareous exoskeletons (Foraminifera) or endoskeletons (Heliozoa and Radiolaria).

The evolution of the pseudopodial protozoans seems to have reached a point of stasis a long time ago. Thompson (1942), after examining shells of fossil Foraminifera, denied that the variations in form had any adaptive significance and considered that no selection had occurred amongst them. Sandon (1968) observed that, while there were broad adaptations to floating, crawling or swimming, little structural basis could be found for these adaptations. The future evolution of the pseudopodial protozoa appears to be severely limited because their poor locomotion, passive defensive construction and sedentary mode of life restrict them to bottom-crawling or floating aqueous environments.

In contrast, the other line of protozoan evolution, that of the zooflagellates and ciliates, has been towards actively moving, predacious organisms of relatively unspecialised external structure. The establishment of predation in this line has involved an increase in size, a further development of locomotor organelles and of organs for the engulfment and digestion of smaller organisms. However, the living zooflagellates are mostly symbiotic organisms belonging to six or seven relict groups which have co-evolved with their hosts (Hanson, 1977). They are both specialised in their habitats and limited in their size, but some free-living members have taken one or other of two ways out of the latter difficulty. In their probable aggregation to form a sponge colony, they have partially alleviated the problem of small size, but evolution then stopped. The most successful strategy was the elaboration of the complex internal organisation of the ciliates, a complexity which is not approached by any other organism, including the highly differentiated ones. Thus, while all the ciliates have special organs for movement, mouths and anuses, some have an oesophagus-like tube, a rectum-like tube and something resembling an alimentary canal (Sandon, 1968), all in a single cell. But once again, the complexity or perfection of ciliate internal structure seems to have run them into an evolutionary blind alley. This is shown by the many ciliates which exist as assemblages of 'syngens' or 'biological species' which, although apparently identical morphologically and biochemically, are completely isolated genetically. Their phenotypes have continued unchanged in at least a dozen different lineages of *Tetrahymena* for possibly more than 500 million years (Nanney, 1982). Sexuality is commonly absent and, although there is very great diversity between syngens in nucleic acid sequence and protein electrophoretic variation, this can be attributed to selectively neutral changes ascribable to the molecular clock (see Section 1.2).

The earliest evolved eukaryotes of which there is any molecular knowledge are the Kinetoplastida (flagellated protozoa). Many of them, such as *Trypanosoma, Leishmania, Crithidia* and *Plasmodium* are now intracellular parasites of mammals, but before that they must have existed as free-living organisms. The structure of their genes seems to be not very different from that of prokaryotic genes in that, unlike other eukaryotes, transcription can occur in a polycistronic manner and, as yet, no intervening sequences have been found in any enzyme-coding genes. One advance over prokaryotes is an increase in the compartmentation of gene expression: for example, most enzymes of the glycolytic pathway are found in glycosomes, which are membrane-bounded organelles akin to the peroxisomes or microbodies of higher eukaryotic cells. Their genes also code for two short sequences of two to ten positively-charged amino acids in the proteins, which are presumed to provide the signal for their importation into glycosomes (Michels, 1986). Sometimes, as with phosphoglycerate kinase, one isoenzyme is directed to the glycosome while the other remains free in the cytoplasm. These isozymes have only 7% difference in sequence as the result of repeated conversion between the duplicate genes (Osinga et al., 1985). They show three characteristic features of gene conversion, a nonreciprocal transfer of blocks of sequence between genes, high homology of the sequence preceding the variable region and distinct boundaries to homologous and variable regions.

Other housekeeping genes in trypanosomes that are present in tandem arrays, such as those for glycerophosphate dehydrogenase and calmodulin, are also kept highly homologous. Calmodulin is encoded by three identical tandemly repeated genes separated by spacers. Each messenger RNA has a 39-nucleotide leader spliced to the 5' end that is not encoded in the calmodulin locus nor in adjacent genomic DNA, but in a separate tandemly repeated DNA segment of about 200 copies (Tschudi et al., 1985). The same leader sequence may be attached to the 5' end of many, if not all, trypanosome RNAs, suggesting that the discontinuous transcription of protein-coding genes is widespread in these protozoans.

The flagellated protozoa, being intracellular parasites of antibody-forming organisms, are subject to immunological attack directed against the surface glycoprotein. As a defence, the parasites use the strategy of antigenic variation, as also do the prokaryotes *Salmonella* and *Borrelia*. However, parasites like *Trypanosoma* can activate individual members of a large set of genes coding for different surface antigens. In one group of such genes, an expressed gene near the chromosome end is converted from a silent basic gene copy, while in a second group a gene becomes activated for expression without duplication or conversion. Alternation of these two gene switches creates a wide variety of antigens (Laurent et al., 1984). In the malarial parasite *Plasmodium*, on the other hand, the antigenic surface protein contains about twenty similar repeating units which rapidly evolve by mutational substitutions combined with unequal crossing-over to give shifting patterns of substitutions (de la Cruz et al., 1987). Perhaps in connection with antigenic protein alteration, a novel mechanism for restoring the reading frame of mutations that produce frame shifts has been reported in *Trypanosoma*. A frame shift mutation in the mitochondrial cytochrome oxidase gene is suppressed by the insertion of four uridine nucleotides that are not encoded in the DNA (Benne et al., 1986). Differences between the two alternative hosts, insects and mammals of *Trypanosoma* in this RNA editing suggest that the process may have originally evolved as a translational regulatory mechanism.

The ciliates are peculiar in that the cell contains a germinal micronucleus and a somatic macronucleus. The macronucleus develops after conjugation or autogamy from the micronucleus by a process of fragmentation and an approximately 400-fold amplification of the DNA.

The ciliate nuclear code also differs from the 'universal' code as found, for example, in mammals. Some codons that are used for protein termination in the usual code, such as the ochre triplet TAA and the amber codon TAG, code for glutamine in *Paramecium*, *Tetrahymena* and *Stylonychia* (Horowitz and Gorovsky, 1985). The ciliate mitochondrial code differs from the nuclear one in that TGA is translated as tryptophan, ATA is used as an initiator codon probably for methionine and TAA serves as a stop codon (Ziaie and Suyama, 1987). These coding differences, together with the fact that conserved genes such as those for histone, actin and ribosomal and transfer RNAs show less homology than the genes of other eukaryotes, may suggest that the ciliates branched off from the eukaryote lineage extremely early, perhaps at a time when the coding system was still evolving. The ribosomal RNA sequences in the ciliates do not, however, indicate a very early origin (Baroin et al., 1988), so the coding characteristics may be derived rather than primitive.

In addition to expressing a high degree of intracellular complexity, the ciliates possess complex cell surface patterns elaborated from units organised around ciliary basal bodies. The patterns are highly conserved although the ciliary proteins, major cytoskeletal proteins, proteins of the oral cytoskeleton and surface membrane proteins differ substantially among species (Seyfert and Willis, 1981). This variability in structural proteins may in part stem from a characteristic of ciliate genes, the presence of multiple start sites of transcription provided by putative TATA sequences (Barahona et al., 1988). The isolation and sequencing of a surface protein gene from *Paramecium* reveals that it has a highly periodic structure, suggesting an origin by multiple duplications of a small basic gene. Its only known homology is with genes coding for surface antigens in trypanosomes (Prat et al., 1986).

The genes of the slime-moulds, both cellular (*Dictyostelium*) and plasmodial (*Physarum*), do not show much change from homologous genes of prokaryotes. About half of the highly repeated sequences of *Physarum* are probably due to the replicative transposition of an 8600 nucleotide long transposon-like repetitive element (Pearston et al., 1985). In this respect, the repeated part of the genome resembles eukaryotes like *Drosophila* and mammals, rather than prokaryotes. This transposon preferentially integrates into sites located within its own sequence, as also does the heat-shock responsive transposable element of *Dictyostelium.*

Some genes unique to eukaryotes, like those for myosin, actin and ubiquitin, have been isolated in probably their most primitive form in

137

Dictyostelium They are closely similar in structure and size to the homologous genes of advanced eukaryotes and generally seem to have arisen by a series of tandem repetitions of quite short modules together with some nucleotide substitutions. However, intervening sequences in *Dictyostelium* genes are infrequent, and when present are relatively short in comparison with the sequences of other organisms. A characteristic and unique feature of all *Dictyostelium* genes so far isolated is the presence in promoters of sequences high in adenine and thymine 5' to the start of transcription initiation (Kimmel and Firtel, 1983). Since adenine polymers can promote bending of DNA, these sequences may be involved in the regulation of transcription.

The rhizopodal protozoa are thought to have arisen about 1000 to 1200 million years ago, after the fungi diverged from the eukaryotic line of descent, but before the divergence of plants and animals. Few of their genes have been isolated but that for myosin I in *Acanthamoeba* contains 23 intervening sequences, nearly half of which are conserved in position in the homologous gene from the rat (Jung et al., 1987). This observation suggests that intervening sequences were present in some genes at least 1000 million years ago and is in accord with the idea that they were in the first genes to arise, rather than being inserted later to give the divided structure of vertebrate genes.

4.5.3 *Fungi*

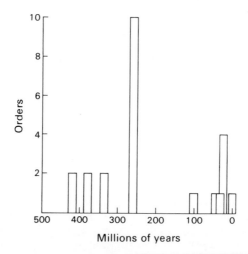

Figure 32 The times of first appearance of living orders of fungi according to the rather poor fossil record (data from Tiffney and Barghoorn, 1974, and Harland et al., 1967). No known order has become extinct except the doubtful Discellales, a proposed order of the Ascomycetes. Three orders of 'Deuteromycetes' (fungi imperfecti) are included.

The lower fungi are generally believed to have had polyphyletic origins in water and to have come to the land as algal parasites (Saville, 1968). A critical stage in their evolution occurred on emergence from the sea, presumably as their hosts adapted to terrestrial habitats. Fungal adaptation to land involved changes in wall structure and in pigment formation to reduce water loss and protect against radiation, as well as modifications for spore dispersal in air. The very novel environment, and the new genetic adaptations required to cope with it, probably led to a burst of evolutionary radiation, eventually giving rise to the main fungal groups (Figure 32). Although most of these critical intermediates in fungal evolution have disappeared, the fossil record indicates that four of the five major classes of fungi appeared approximately synchronously in the Carboniferous (Tiffney and Barghoorn, 1974).

Their subsequent evolution was determined in large part by their restricted mode of nutrition by absorption through branched systems of multinucleate tubes embedded in their food source. One line led to more highly specialised and host-specific obligate parasites, while the other acquired the ability to grow first on both the live and dead host and, later, on the dead host alone (saprophytism). Still later, there occurred a large increase in the number and diversity of some of the fungi during the Tertiary period which has been attributed to the appearance of the angiosperms with their extensive leaf areas.

Judging from the measured amount of DNA in fungal cells and estimates of genome size by reassociation kinetics (Dusenberg, 1975), the fungal genome is only two to ten times larger than the average bacterial genome. There does not seem to be much relationship between genome size and evolutionary complexity because the earlier and later evolved fungi have similar DNA contents. Ullrich and Raper (1977) considered genetic mechanisms in the fungi and concluded that their evolution had occurred over a protracted period of time and that generally in the fungi evolution was very slow and evolutionary potential limited. Despite their morphological conservatism, the degree of DNA sequence divergence between species seems to be greater than that observed in other groups of eukaryotes (Vilgalys and Johnson, 1987).

Unlike the algae, the fungi appear to have evolved a mode of life which, while being very successful in a limited environmental context, is too specialised to form a basis for future novel evolutionary advance. The major feature of the fungi which limits their evolutionary potential is their dependence upon water, always for vegetative growth and often for reproduction. As a consequence, they are completely reliant on other organisms and must live by the parasitic or the, probably derived, saprophytic style of nutrition.

Most knowledge of gene structure in fungi comes from the yeast, *Saccharomyces cerevisiae*, but in this respect unicellular fungi (yeasts) seem to differ from the great variety of multicellular (filamentous) fungi. As shown in Table 12, page 113, whereas in most yeasts genes tend to lack intervening sequences, they are quite common in the genes of filamentous fungi like *Neurospora* and *Aspergillus*. In this connection it has been pointed out that only 17 genes of *Saccharomyces*, out of several hundred that have been sequenced, contain intervening sequences, usually only one. This loss of intervening sequences has been attributed by Fink (1987) to frequent reverse transcription of functional messenger RNAs, followed by integration via homologous recombination to replace original split genes with processed derivative genes. As shown by the results of transformation, yeast is atypical among fungi in commonly requiring sequence homology for integration to occur; the transformation of filamentous fungi, in contrast, frequently gives nonhomologous and multiple recombination events resembling those occurring during animal transformation.

The majority of intervening sequences seem to be incapable of coding for proteins, but some of the mitochondrial introns of fungi contain long open reading frames that in yeast may code for a protein active in gene conversion. The result is the integration of the intron sequence within all previously intron-free copies of the gene (Jacquier and Dujon, 1985). A similar open reading frame in the introns of a *Neurospora* mitochondrial gene appears to code for reverse transcriptase, perhaps to allow the introns occasionally to move about the genome (Michel and Lang, 1985). This is a system reminiscent of an intervening sequence in the cytochrome C oxidase gene of *Podospora anserina*, which when excised becomes the mobile infectious agent causing senescence (Osiewacz and Esser, 1984).

There is also a difference between unicellular and multicellular eukaryotes in the details of pre-messenger RNA splicing. The yeasts, *Saccharomyces* and *Schizosaccharomyces*, have a stringent mechanism for splicing, as opposed to the greater flexibility in splicing patterns in higher eukaryotes (Mertins and Gallwitz, 1987). This finding fits the model proposed for the selective advantage of intervening sequences, that of transcriptional variation, since the complex splicing patterns during the generation of different messenger RNAs from one gene transcript in multicellular eukaryotes have not been observed in unicellular ones. Although in unicellular fungi it may be desirable for a single gene to code, say, for an intracellular and extracellular protein, this may be achieved by differential transcription initiation. A single yeast gene, for example, makes two forms of invertase, a secreted glycosylated form regulated by glucose repression, and an intracellular nonglycosylated enzyme produced constitutively. Glucose acts to regulate the point at which the initiation of transcription takes place (Carlson and Botstein, 1982).

The position of intervening sequences in genes of the filamentous fungi is probably indicative of a primordial structure. Thus the gene for actin in *Thermomyces lanerginosus* has retained intervening sequences at several positions that have not been found to occur simultaneously in any other evolutionary group (Wildeman, 1988). The ubiquity of actin throughout the eukaryotes is evidence that the actin gene arose before the divergence of eukaryotes and of the probability that intron loss has occurred in different patterns. A similar interpretation of fungal intron position has been made for the gene for triosephosphate isomerase from *Aspergillus nidulans* (McKnight et al., 1986).

Among the genes that the fungi probably evolved after they first arose are ones encoding enzymes active in the penetration of living plant cells or the degradation of dead cells. For example, the major cellobiohydrolase of *Trichoderma reesi* has a structure similar to that of bacterial cellulases, although there is practically no amino acid sequence homology and, unlike the bacterial cellulases, the fungal enzyme can hydrolyse crystalline cellulose in the absence of endoglucanases (Teeri et al., 1987). Both the genes for the cellulase and those for the endoglucanase seem to be constructed on a modular basis comprising a region for the binding of cellulases to their substrates and an acid catalytic domain. Similarly, the endoglucanase of the white rot fungus *Schizophyllum commune*, shows sequence homology with the active site of lysozymes (Yaguchi et al., 1983). There is, however, no evidence in any of these genes of assembly by module exchange via recombination within intervening sequences.

4.5.4 *Plants*

Higher eukaryotes clearly split at the unicellular protistan level, probably from photosynthetic algal-like ancestors with cellulose walls. The plant line retained the cellulose wall surrounding the membrane, leaving the animal line of evolution to develop the extensive potentialities of cells with flexible boundary membranes. The advantage to the unicellular plant of a rigid noncompressible wall is that it can maintain a high internal osmotic pressure without having to develop the complicated osmo-regulation of the naked animal cell. However, rigid walls have meant that, when the multicellular condition was reached, the evolving plant was committed to a fixed mode of life and an existence dependent on a minimum of immediately adjacent inputs, mainly water, minerals and light. The requirement for growth in an inextensible body, as well as most of the cell differentiation programmes associated with a colonisation of the land, were met by the elaboration of a few developmental strategies that originated in the algae over their several hundred million years of existence.

The line of biological advance from the algae to the angiosperms was initially concerned with the development of cell and tissue differentiation, and later with the origin of vegetative and particular sexual organs. The first main phase of evolution, completed at least 400 million years ago, involved the adaptation of algal progenitors resembling *Nitella* to life on the land. This shift from an aqueous to a part-aerial environment was made possible by the acquisition of means of withstanding the relatively dessicating conditions of the land. This led to the evolution of water-conducting tissues, of protection for sex organs and of a water-impervious cuticle to minimise water loss. Some of the intermediates in this process are still extant. Such are possibly the bryophytes where the primitive method of fertilisation and the lack of vascular tissue has limited their size, distribution and further evolution, and the pteridophytes whose future is blocked by their requirements for water in fertilisation and in the maintenance of the gametophyte. An independence of water in the sexual process, for which there was no algal starting point, apparently took about 100 million years. It was then accomplished by a progressive reduction in

141

the prothallus stage of the pteridophytes. Subsequently, the advantages for photosynthesis and reproductive dispersal of the projection of the vegetative plant body into the air required the evolution of food-conducting cells and of cells giving mechanical support to the aerial plant parts.

Later plant evolution has involved little apparent increase in genetic information or in basic biochemistry, nor has it involved a great extension of pre-existing systems of differentiation and morphogenesis. Even the most important synthetic process of photosynthesis has remained virtually unchanged from that of the algae. Their evolutionary processes seem to have been primarily concerned with the utilisation of their metabolic byproducts as repellants and attractants, and the perfection of their reproductive systems. The latter reach their peak in the angiosperms (Figure 33) where there is an enormous variety, mostly variations on the theme of sepals, petals, stamens and carpels, and with highly evolved mechanisms of seed dispersal.

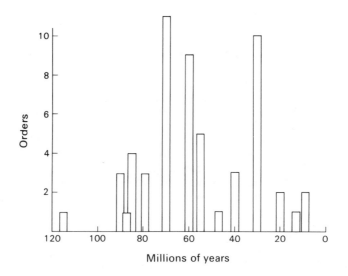

Figure 33 The times of first appearance of living orders of dicotyledonous plants according to the fossil pollen record (data from Muller, 1981). Plant identification by pollen is generally considered more accurate than that by leaf remains (Dilcher, 1974). These orders are treated as families in Harland et al. (1967).

With the rise of the angiosperms at least 100 million years ago in the Cretaceous, the nature of the earth's flora changed rapidly. The tree-ferns

142

and gymnosperms which had formed the dominant vegetation were soon displaced and, by 30 million years ago, 90% of the angiosperm orders (or families) had already evolved. Some of them, like the Gramineae, Orchidaceae, Leguminosae and Compositae have proved extraordinarily successful, each producing about 15 000 species (Ross, 1966). Neither the reason for the success of the angiosperms, nor that for the specific families mentioned, is at all clear. A variety of features, chemical, anatomical, morphological and reproductive, has been proposed to account for angiosperm advantage, but all are subject to objections or exceptions. A few trends in angiosperm evolution seem to be continuations of earlier established progressions. One frequent direction is from trees to arborescent shrubs to smaller shrubs to perennial herbs. Cronquist (1968) considers that this is brought about by neoteny resulting in the onset of flowering in successively earlier ontogenetic stages. During this process, there is generally a decrease in the activity or presence of the cambium, so the reverse tendency from herbaceous to woody habit is rarer and is dependent on the cambium being retained in becoming herbs. Another continuing trend is for deciduous angiospermous trees like birch and alder to replace gymnosperms in northern latitudes. The rate of speciation in gymnosperms is much lower than in angiosperms, which Levin and Wilson (1976) attribute to the open breeding system and rather poor seed dispersal of gymnosperms. There are also directional trends that occur in the flower, probably concerned with increased protection, especially to the ovary and zygote, and with enhancing processes of fertilisation. However, many floral features characterising species, genera and families appear to be alternative states of a structure, with approximately equal advantage.

Taken as a whole, the mode of life of the higher plants is more resistant to environmental disturbance than that of animals. Plants are less susceptible to radiation, noxious gases and fluctuations in water availability and temperature, the main components of catastrophic environmental change. However, their potentialities for further evolution in such circumstances seem rather limited because of their immobility and restricted differentiation patterns. The various adaptations currently existing give some idea of the potentialities of the plant design. Thus, parasitism exists and is a future possible evolutionary pathway, although it is only other plants that are parasitised. Saprophytism, on the other hand, is rare presumably because plants have difficulty in evolving the necessary enzymes and absorptive systems for utilising decaying organic matter, although it is abundant. As shown by the Droseraceae, among others, many plants in appropriate circumstances could become carnivorous. Some existing insect-catching plants have organs sensitive to touch and movable trapping devices; and they can secrete formic acid and a peptidase for the digestion of insects. This adaptive complex must represent the apex of plant evolution. Finally, plants could return to the sea from which their ancestors came. *Zostera* is such a flowering plant which now lives totally submerged below the low tide level of the sea.

The genes of vascular plants differ in origin according to whether they are from the nucleus, chloroplast or mitochondrion. The chloroplast genome is usually from 100 000 to 180 000 base-pairs in size, with its known genes being concerned with the transcription-translation system or with photosynthetic electron transport. A comparison of the complete nucleotide sequences of the chloroplast DNA from the bryophyte *Marchantia polymorpha* and the flowering plant *Nicotiana tabacum* (Ohyama et al., 1988) indicates that most evolutionary events in the chloroplast were completed at least 400 million years ago, the palaeontological date for the origin of the bryophytes. The chloroplast genome has many prokaryotic features such as a polycistronic organisation of transcription units, prokaryotic-like promoters and similar terminators to prokaryotes (Boyer and Mallet, 1986). However, unlike prokaryotic genes, many genes in the chloroplast contain intervening sequences of various types which, by their structure and position, provide a basis for assessing the evolutionary history of chloroplast genes.

The plant nucleus contains two independent genes each coding for a distinct glyceraldehyde-3-phosphate dehydrogenase. The one active in the chloroplast is involved in photosynthetic carbon fixation, while that of the cytosol (cytoplasm) functions in glycolysis. A comparison of nucleotide sequences indicates that the two genes evolved from different lineages that diverged about 1700 million years ago (Shih et al., 1986). This finding supports the proposition that the chloroplast has evolved from an endosymbiotic prokaryote and that nuclear genes encoding chloroplast-specific functions have been transferred from the endosymbiotic genome to the nucleus. Comparisons of amino acid sequences suggest that the chloroplast enzyme is related to that of thermophilic bacteria, whereas the cytosolic enzyme resembles that found in mesophilic prokaryotes. Furthermore, the nuclear-located gene for the chloroplast-functioning glyceraldehyde-3-phosphate dehydrogenese from maize contains three intervening sequences, whereas the gene from the nematode *Caenorhabditis* contains two intervening sequences. The first two intervening sequences of maize are located in the sequence for the transit peptide, but the third separates the NAD-binding and catalytic domains of the enzyme at precisely the same position as the first intervening sequence of the nematode gene (Quigley et al., 1988). This suggests that this intervening sequence was present in the ancestral gene before its division into photosynthetic and glycolytic representatives, i.e. before the separation of prokaryotes and eukaryotes.

Some of the changes needed to adapt a prokaryotic gene to the chloroplast are revealed by a detailed comparison of the nuclear-coded small subunit genes for ribulose bisphosphate carboxylase from the potato with the homologous gene from cyanobacteria (Wolter et al., 1988). The plant gene consists of a set of short coding sequences separated by intervening sequences. The first coding sequence is not present in the prokaryotic gene and codes for a transit peptide, the second is homologous to part of the prokaryotic gene, the third has no homology to prokaryotic sequences and probably makes a peptide that assists in the interaction of the large and small subunits of the enzyme, and the fourth again shows homology to prokaryotic small subunit genes. The evolution of this gene in plants is thought to have begun with its transfer from an endosymbiotic prokaryote to a position in the eukaryotic nucleus. This event necessitated the addition of a domain coding for a transit peptide as well as a signal peptide for coordinating the expression and assembly of the products of genes located in separate compartments. The separation of these functional domains by intervening sequences suggests that they were assembled into one gene by recombination within these sequences. Thereafter, the intervening sequences may have been progressively reduced in number during plant evolution. The small subunit genes of the Solanaceae (*Petunia, Nicotiana, Solanum*) contain three intervening sequences, while other dicotyledonous plants have two, and monocotyledonous plants (*Triticum, Zea* and *Lemna*) have only one intervening sequence. The archaic nature of the gene with three intervening sequences is indicated by the fact that this gene in the primitive green alga *Chlamydomonas* also has three.

In both monocotyledonous and dicotyledonous plants, the numbers and border sequences of intervening sequences are similar, yet genes of monocotyledonous plants that contain intervening sequences are usually poorly expressed, if at all, in the cells of dicotyledonous plants (Hanley and Schuler, 1988). The inefficient expression in tobacco of genes for ribulose bisphosphate carboxylase of wheat and alcohol dehydrogenase of maize appears to result from an inability of the dicotyledonous splicing machinery to recognise the 5' splice sites of monocotyledonous intervening sequences. The variety of splicing signals that exist in the intervening sequences of plants may mean that the evolution of new genes by module exchange and assembly is of only limited occurrence. A fairly recently evolved gene in legumes for nodulin, a protein involved in symbiotic nitrogen fixation by *Rhizobium*, has evolved by internal duplication (Katinakis and Verma, 1985). The gene's structure suggests that the duplications were of an inserted sequence coding for a hydrophobic domain of membrane-like character. As far as we know at present, module recombination in the generation of plant genes was confined to primordial genes that were present before the flowering plants evolved.

So far, no examples of alternative splicing have been found in the transcriptional expression of flowering plant genes. Organs of higher plants are composed of only ten basic tissue types, so the production of proteins adapted to particular tissues is not the problem that it is in animal organs. Therefore, the limited requirement for differential expression in plant genes, as in the genes for filamentous fungi, can be satisfied by an initiation of transcription at variable sites. Thus the expression of the gene for 5-enolpyruvylshikimate-3-phosphate synthase in *Petunia* starts nearly equally at several initiation sites in leaves, but predominately at a single site in petals (Gasser et al., 1988).

4.5.5. *Invertebrates*

The invertebrates (Figure 34), although very heterogeneous and of diverse phylogenetic origins, probably comprise more than 90% of all known species of animals. In their evolution, they had first to form cellular aggregates, then initiate cell differentiation programmes to provide a division of function in the cellular mass. Later some of them formed a two-layered animal by the passage of some cells from the surface to the interior. The next stage in evolutionary advance was probably the formation of a mesoderm for the development of organs and the acquisition of blood circulation and a nerve network. Most invertebrates are marine animals, and all these evolutionary steps had been completed and all the marine phyla established by the Ordovician; since the end of the Permian, the invertebrates have been in evolutionary stasis with respect to both phyla and classes.

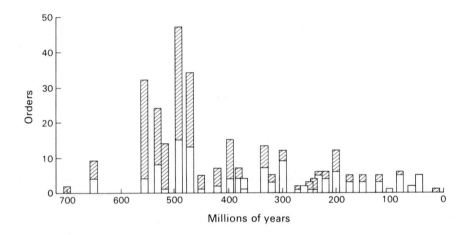

Figure 34 The time of first appearance of orders of marine invertebrates (about 22 phyla comprising Porifera to Hemichordata, inclusive) according to the fossil record (data from Sepkoski, 1982). Among those marine orders labelled extinct, a few may still persist on land. Open bars are living orders; hatched bars are extinct orders.

The two groups of organisms representing the early stages in invertebrate advance are the Porifera (sponges) and the Coelenterata (polyps, jellyfish, anemones etc.), the latter known from the Ediacaran fauna of 670 million years ago and the former from the Middle Cambrian. The sponges achieved a rather loosely knit aggregation of cells, but developed neither a mouth nor a digestive cavity. They feed by means of water canals, and the only evolution they appear to have undergone since their radically symmetrical bodies were established has been the appearance of new sponges, which were more efficient in creating a water current. The sponges apparently started with a limited evolutionary potential which was soon used up.

The coelenterates, on the other hand, evolved a mouth leading to a gastrovascular cavity. They show great plasticity in matters relating to food capture, the main evolutionary trends being towards various kinds of nematocyst-bearing tentacles around the mouth and in the form and internal structure of their bodies. Yet they have never developed other than a simple organisation and a low level of tissue differentiation. Hanson (1977) remarked that their postulated origin from ancestral zooflagellate cells with little organisational and developmental potential has doomed the coelenterates to a restricted evolutionary development that not even colony formation could overcome.

Four of the five existing major phyla of worms, Platyhelminthes (flat-worms), Rhyncocoela (ribbon-worms), Aschelminthes (nematodes, rotifers etc.) and Acanthocephala (spiny-headed worms) are rather simply organised and many of them live as parasites. The earlier forms were in a sense pre-adapted to parasitism because of their body shape and rather poor digestive systems. Frequently associated with parasitism is a high degree of specialisation combined with structural degeneration. The most evolutionarily versatile group is the nematodes, probably mainly because of their highly developed ability to survive adverse conditions by arresting their development and by cryptobiosis, a dormant state with almost imperceptible metabolism. The remaining phylum of Annelida (marine worms and earthworms) is more highly organised and some of its members have developed, or have begun to develop, nearly every type of organ found in much higher organisms. There may be a gut with a mouth, jaws and anus, a coelom, gills, circulating system, nephridia to remove nitrogenous byproducts, nerve ganglia and a primitive brain (Romer, 1968). However, functionally similar organs in vertebrates are believed not to be derived from the annelids.

Except for the insects, the Mollusca have the largest number of species among invertebrates and both the bivalves and the gastropods have existed with little change since the Cambrian. Early in their history, however, the molluscs developed a mantle, over the surface of which they secreted a hard shell of calcium carbonate. Although this gave them a good defensive construction, they lost much of their evolutionary potential since they became nearly or wholly sedentary and abandoned the advantages of segmentation. Much later, probably during the Ordovician, a section of the molluscs made an attempt to alter their whole pattern of existence by becoming aggressive predators. These are represented by the modern Cephalopoda (squids, octopuses and cuttle fish). In the process they evolved such innovations as a jet-propulsion system for rapid movement, prehensile tentacles for the capture of prey, and eyes with a visual acuity achieved elsewhere only by the higher vertebrates. But their basic structural pattern, in particular deficiencies of their digestive system and nephridia, combined with the low oxygen-carrying capacity of their haemocyanin respiratory pigment, have limited their success. They may be capable of still further evolution, and their presence shows the possibilities latent in molluscan organisation. Nevertheless, the majority of the molluscs, although they are as abundant and diverse now as they

147

have ever been, have never been able to produce efficient terrestrial forms; they became evolutionarily static a long time ago.

The filter-feeding invertebrates (bryozoa, brachiopods, and the more primitive echinoderms and crinoids) are thought to represent a few surviving end-groups of a broad radiation of sessile forms early in metazoan evolution (Romer, 1968). The bryozoa are tentacled bloodless invertebrates which have continued practically unchanged since the Lower Ordovician, except for the extinction of some orders. The brachiopods, although very abundant in the Palaeozoic, started to decline in the Carboniferous, perhaps from competition with the rapidly evolving bivalve molluscs, and now are reduced to only about 160 species. However, some of the early genera like *Lingula* and *Crania* have persisted from the Silurian to the present, for unknown reasons. The echinoderms probably started their evolutionary history from bilateral ancestors in the Proterozoic or early Cambrian, but adopted the radial body-plan as an adaptation to a sessile existence. During this change, they lost any specialised sense organs and brain that their ancestors might have had, and the scope of their evolution has been largely determined by their radial structure, water-vascular system and dermal endoskeleton (Hickman et al., 1979). The evolution of the echinoderms has been very slow, especially in the attached forms. The most notable trends in the mobile lines were the development of hydraulic suction in the tube feet of asteroids (starfish) and the jointed arms for locomotion in the ophiuroids (brittle stars). But the echinoderms have not advanced for perhaps 100 million years, nor have they given rise to other major groups.

In terms of biological 'success', as measured in numbers of species and of individuals, the arthropods are the most successful of all animals. Of the approximately one and a half million contemporary species of eukaryotes, including algae, fungi and protozoa, estimated to exist (Shorrocks, 1978), three-fifths or about 0.9 million are insects. The earliest known probable insect is *Eopterum* from the Devonian, and by the end of the Carboniferous ten or eleven orders had become established (Figure 35). However, according to Ross (1966), no new order of insects has evolved subsequently except perhaps for two small orders that live as ectoparasites on land vertebrates. A large burst of radiation occurred in the Permian about the time land plants were also developing rapidly. There was extinction of many orders in the Mesozoic and, essentially, modern insects had appeared by the Tertiary (Riek, 1970). The insects have, as a result of their small size, rapid ontogenesis and high fecundity, become one of the most successful life-patterns that have evolved, but their limitations are inherent in their morphological construction. Because of their tracheal respiration and the necessity for moulting, they cannot become larger than they are now. In turn, their small size prevents the development of a temperature-regulating system and limits the size to which their brain can grow. Some of them, however, have partly overcome these limitations by becoming social insects. The bees, ants, wasps and termites adopted a colonial form of behaviour by the Cretaceous to middle Tertiary, enabling them effectively to defend themselves, build complex nest structures and

148

occupy a larger range of territory (Wilson, 1971). The persistence of the ants practically unchanged and without reduction in species number for at least 50 million years attests to the success of the colonial strategy. Some of the social insects like the termites and certain ants cultivate fungi, other ants keep individuals of a related species as slaves and yet other ants raise and tend aphids for their sugary excretions. The insects could be reasonably regarded as specialised end-products of evolution were it not for the possibility that they could re-evolve from one of their intermediate stages of development. According to De Beer (1958), the insects originated from the larval form of a millipede ancestor, and the female glow-worm, which remains in the larval condition, is a living example of the possibilities of neoteny.

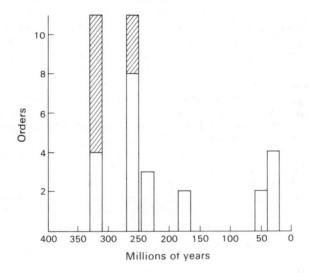

Figure 35 The times of first appearance of orders of insects according to the fossil record (data from Riek, 1970 and Carpenter, 1977). Of the four orders shown as originating in the Tertiary, the Mantodea and Siphonaptera are poorly represented as fossils, the Embioptera are shown as originating in the Permian by Crowson et al., (1967) and Riek (1970), and some authors regard the Strepsiptera as a superfamily of the order Coleoptera. Open bars are living orders; hatched bars are extinct orders.

The lesser classes, the crustaceans, centipedes, millipedes and arachnids, comprise only about 7% of all arthropodan species. The crustaceans suffer

149

in particular from their chitinous armour which necessitates repeated moulting, limitations in size and specialised but weak appendages. On the other hand, the chitinous exoskeleton has provided an adequate barrier to dessication, which, combined with tracheal respiration, has enabled some species to live on land. Few of the Crustaceae have successfully managed the transition from water to the land, but the arachnids (spiders, scorpions etc.) have done so very successfully (more than 50 000 contemporary species). Evolution in the Crustaceae has been very slow indeed, as exemplified by two species of *Australoniscus*, the woodlouse. *A. alticolus* in Nepal and *A. springetti* in Australia have been separated for 135 to 140 million years by the breakup of Gondwanaland yet they differ only by the endopodite end of the first male pleopod which is straight in *A. springetti* and bent into a hook in *A. alticolus* (Vandel, 1972).

It is generally believed, on the basis of fundamental similarity of structure, that the chordates arose from the larva of ascidians (sea squirts), although not from larva of the modern type where the free-swimming stage has been reduced to a matter of hours or a few days (Barrington, 1967). The adult ascidians are now very degenerate, being in general permanently attached animals with tough surrounding 'tunics', with elaborate filtration systems, colonial existence and other forms of specialisation. There seems little chance that the chordates could again arise from the ascidians or that ascidians themselves will evolve further.

The invertebrates, particularly *Drosophila*, provide an opportunity to examine the origin and evolution of some genes involved in development. Some of these developmental genes probably represent early evolved genes in an insect's genome, marking a transition from the largely cell-intrinsic programming of invertebrate worms like the nematode and leech to the more advanced mechanism of inductive interaction in arthropods and vertebrates.

The development of diverse body segments requires the action of genes controlling segmentation, as well as homoeotic genes to establish the morphological pathways appropriate to each segment. Three classes of segmentation genes are known; gap genes that divide the embryo into regions, pair rule genes that divide the regions into individual segments and segment polarity genes that control the morphogenesis of the segments. These genes encode proteins in which certain recurrent motifs appear. The homoeo domain is a sequence resembling DNA-binding proteins in bacteria and sequences involved in transcriptional regulation in some yeast genes. A sequence giving a 'zinc finger' structure (see below) in gap genes may also be involved in transcriptional regulation. An unrelated structure, the paired box region, is found encoded by several genes controlling embryonic polarity.

One class of developmental control gene is represented in *Drosophila* by Krüppel, a gene involved in the proper segmentation of the embryo, mutants of which lack the thoracic and first five anterior abdominal segments. This gene codes for repeated sequences homologous to that for one of the transcription factors of *Xenopus* (Rosenberg et al., 1986). The protein is thought to fold into a metal-binding core with finger-like projections that make contact with nucleic acids. Other genes for similar metal-binding and nucleic-acid-binding structures have been found in bird and mammalian DNA, suggesting that they may also control gene expression during vertebrate development. Since the presumed finger-coding regions in the gene are separated by intervening sequences (Tso et al., 1986), it is possible that the gene evolved by successive duplications of a small DNA-binding molecule.

Two of the known invertebrate homoeotic loci, *Notch* in *Drosophila* and *lin-12* (for lineage abnormal) in *Caenorhabditis*, encode proteins that contain characteristic signal and transmembrane sequences as well as an extracellular domain that is homologous to epidermal growth factor (EGF) of mammals. EGF causes proliferation of a variety of cells through its binding to membrane receptors found in their surfaces. The pleiotropic effects of EGF suggest that it behaves as a differentiation switch acting nonspecifically since it is also found in other mammalian proteins like urokinase, tissue plasminogen activator, low density lipoprotein receptor and some blood-clotting factors (Bender, 1985). Each EGF-like repeat in the *Notch* locus is not encoded by a separate RNA coding region as it is in the mammalian genes. Rather, it consists of clusters of EGF-like repeats in *Notch* separated by intervening sequences which are not found in conserved positions, nor are they all in the same phase. For these reasons, it is likely that *Notch* evolved by tandem duplication of one or a few original repeats, followed by the insertion of intervening sequences. In mammals, on the other hand, the encoding of EGF-like sequences by separate RNA coding regions suggests the occurrence of exchange between EGF sequences in different proteins (Kidd et al., 1986).

As compared with vertebrates, flowering plants and filamentous fungi, the invertebrates have fewer intervening sequences in their genes (Table 12, page 113). However, in *Drosophila*, genes coding for developmental functions seem to possess genes that are more frequently split by intervening sequences than are genes concerned with metabolic or structural functions (Table 14). In both classes, the presence of intervening sequences allows alternative RNA splicing, leading to the encoding of more than one protein from a single gene. Thus the gene in *Drosophila* for the major structural protein, myosin heavy chain, contains three intervening sequences. At least four different mature transcripts are produced, differing by the inclusion or exclusion of a coding sequence (exon) and by a choice of polyadenylation site (Bernstein et al., 1986). Such transcriptional differences give myosin heavy chains that are specific to developmental stage or tissue. A very similar pattern of alternative splicing is shown by *Ultrabithorax*, a homoeotic gene for segment identity in *Drosophila*. At least five distinct splicing products, which differ in their small internal coding sequences and in alternate sites of polyadenylation, are formed in response to the requirements of different tissues (O'Connor et al., 1988).

Some other developmental genes of *Drosophila* seem to have arisen by the adaptation and specification of earlier-evolved metabolic genes. The maternal effect gene, *snake*, which is responsible for the establishment of

TABLE 14

Gene structure in *Drosophila*. A table showing that intervening sequences are, on average, twice as frequent in developmental genes as in metabolic genes.

A. Developmental genes

Gene	Function	No of Intervening Sequences
Zeste	Regulates other loci	2
Krüppel	Abdominal segmentation	1
Decapentaplegic	Dorsal-ventral determination	2
Notch	Ectoderm differentiation	8
Deformed	Homoeotic selector in head region	4
Antennapedia	Homoeotic selector in antennae	7
Ultrabithorax	Homoeotic gene for segment identity	4
Engrailed	Regulation of early embryo segmentation	2
Glued	Organisation of visual system	4
Fushi tarazu	Formation of alternate segments	1

B. Metabolic and structural genes

Gene	Function	No of Intervening Sequences
Glucose-6-phosphate dehydrogenase		3
ß-tubulin		3
Vittelogenin		1
Alcohol dehydrogenase		2
Actin		1
Trypsin		0
Cu-Zn superoxide dismutase		1
Heat-shock protein (23-27kd.)		0
Myosin heavy chain		3
α-amylase		0
Calmodulin		3

the dorsal-ventral axis in *Drosophila* embryos, appears to encode a serine protease (De Lotto and Spierer, 1986). The protein has 29 to 35% homology with other proteases, and like them it possesses the usual pattern of disulphide bridges, hydrophobic signal peptide and calcium-binding domain. It is not clear what modifications for developmental purposes have been made by the insects, but the basic structure of the gene was already established in the bacteria (Olson et al., 1970). Where a developmental pathway depends on cell-to-cell communication to provide positional information, genes are required both for the synthesis of the developmental signal and for receptors to receive that signal. A gene of the latter type is *sevenless* in *Drosophila*, which is needed for the development of one of the photoreceptor cells of the compound eye. The gene encodes a transmembrane protein of which the carboxy-terminal region is homologous with the tyrosine kinase domains encoded by several viral oncogenes and by hormone receptor genes like that for epidermal growth factor (Hafen et al., 1987). The phosphorylation of tyrosine is thought to be a means of transmitting growth-stimulatory signals to the cell interior. Two other genes in *Drosophila* involved in the selection of developmental pathways, *Notch* and *decapentaplegic*, encode proteins that are homologous to ones concerned with hormonal regulation.

The data presently available on the origin of genes concerned with development in *Drosophila*, although very limited, suggest that at least some of them have been made by the modification of genes active in metabolism; there is no indication so far that any of them had a unique origin.

4.5.6 *Vertebrates*

The major taxa of vertebrates, comprising the fishes, amphibians, reptiles, birds and mammals, have a branching order based on 5S ribosomal RNA sequences, which is in good accordance with the classical (palaeontological) view. Although they are monophyletic in origin, vertebrate classes have differed considerably in their frequencies of ordinal origins and extinctions.

Fishes

The various classes and subclasses of fishes, cartilaginous, lobe-finned, lung and ray-finned fishes, existed as distinct, anatomically specialised, groups in the Devonian, and they have changed very little since. The elasmobranchs (sharks, skates and rays) were the first to become dominant in the ocean, but declined drastically by the Permian, for unknown reasons. They revived to some extent in the Mesozoic, evolving into successful animals with streamlined elongate bodies and aggressive predacious habits on the one hand, and specialised bottom-dwelling sharks like skates and rays on the other. However, the mode of swimming of the typical shark seems less efficient than that of the advanced bony fishes. This is because the shark's elongated shape means that the waves of muscular contraction passing down the body are relatively short. Also the asymmetrical tail with a larger dorsal than ventral lobe tends to drive the body downward, a tendency that is partly corrected by a flattened head (Young, 1950). Much of the shark's ocean environment was taken over by the bony fishes which began to move from fresh to salt water in the Mesozoic era. Subsequently, these ray-finned fishes underwent marked adaptive radiation, and one superorder of fishes succeeded another in replacement evolution. The chondrosteans were dominant during the late Palaeozoic but were replaced by the holosteans in the early Mesozoic and replaced in turn by the teleosteans (Figure 36) in the late Mesozoic and Cenozoic (Colbert, 1955). These almost complete replacements were not brought about by climatic changes as far as is known, but mainly by the evolution of increasingly effective designs for swimming. The improvements involved changes in the caudal, dorsal and paired fins, increased ossification of the skeleton and structural reduction of the scales.

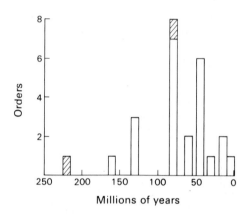

Figure 36 The times of first appearance of orders of teleosts, the most recent subclass of fishes of the class Actinopterygii, according to the fossil record (data from Siegfried, 1954 and Romer, 1966). Open bars are living orders; hatched bars are extinct orders.

The fishes provide one of the few examples of a partial reversal of an evolutionary trend, leading to an escape from specialisation. The ocean sunfishes of the family Molidae evolved from conventional fish, partly by a suppression of the *lateralis* musculature and a great compensatory development of the erector and depressor muscles of the dorsal and anal fins (Raven, 1940). As a consequence, the sunfish are large, oval animals which often bask on the surface of the water. In one species, *Ranzania truncata*, however, the body has become secondarily elongated and much smaller, to approach again that of an ordinary fish.

With the evolution of the teleost fishes, perhaps the most varied, and certainly the most numerous, of the vertebrates became established. They may now have reached their adaptive peak, or they may still be evolving in their aqueous medium, but they are very unlikely ever to leave it.

Amphibians and reptiles
The amphibians were the third group of organisms, after the plants and insects, to transfer from the water to the land. The transition was partly successful, as the amphibians changed from gill- to lung-breathing and developed limbs for locomotion on land. However, other characteristics retained from their freshwater ancestors, such as fertilisation mechanisms, unprotected embryos, cutaneous respiration and lack of thermoregulation, limits them to wet or moist environments. Despite the evolution of several ingenious systems to avoid the necessity for aquatic reproduction, such as rearing the young in pouches on the back or in the stomach cavity, none of these devices were notably successful. Most of the evolution of amphibians and reptiles occurred between the Carboniferous and the Jurassic, and there has been little evolution since (Figure 37). The amphibia now maintain themselves rather precariously on the wet margins between the water dominated by predatory fishes and the land controlled by reptiles and mammals.

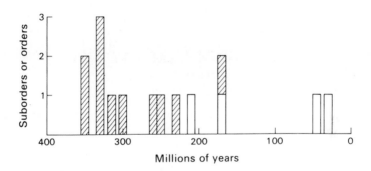

Figure 37 The times of first appearance of suborders, or in their absence orders, of amphibians, according to the fossil record (data from Romer, 1966 and Goin et al., 1978). The Tertiary groups are Apoda and Neobatrachia. Open bars are living orders; hatched bars are extinct orders.

The transition from water to land, which was initiated for the vertebrates by the amphibians, was completed by the reptiles. They improved on the amphibian design by evolving means of reproducing on land, ways of conserving body water, and behavioural patterns providing a degree of body temperature regulation. They reached their peak in number of orders and numbers of individuals in the early Mesozoic (Figure 38) but, apart from the descendant birds and mammals, only remnants of the class, small in size and mostly evolutionarily specialised, remain today. The oldest among them are probably the turtles and the crocodiles, both of which have barely survived, the former because of their protective shells and the latter by returning to the water as large aggressive reptiles. Lizards appeared in the Triassic and are today almost entirely confined to the tropics and subtropics where they lead a modest existence based on an insectivorous life and secretive behaviour. An apparently better design was the snake form, characterised by elongation and flexibility and a loss of limbs which has occurred in members of about seven of the 20 or so families of lizards. The snakes were able to colonise latitudes well beyond the range of lizards but they are the most specialised of living reptiles, and their reductions in structure restrict their capacity for novel evolutionary advance.

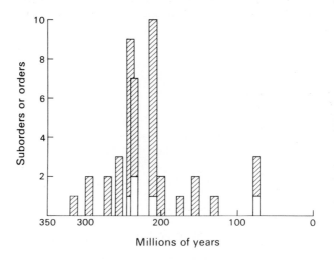

Figure 38 The times of first appearance of suborders, or in their absence orders, of reptiles according to the fossil record (data from Romer, 1966 and Goin et al., 1978). Open bars are living orders; hatched bars are extinct orders.

Birds

The birds represent one of the numerous groups of organisms which have become 'overspecialised'; their skulls, wings, feathers, feet and bone structures form such a complex of specialisations that they have few possibilities for further evolution. They are further restricted by their retention of the reptilian shelled egg which has prevented new paths of evolution by way of neoteny, or the birth of young at an advanced stage of development, and it provides a life stage that is very vulnerable to predation and to the synthetic chemicals used in plant control.

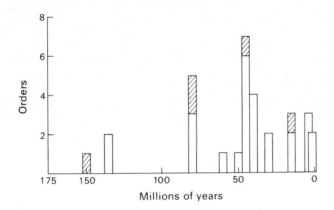

Figure 39 The times of first appearance of orders of birds, according to the rather poor fossil record (data from Romer, 1966 and Brodkorb, 1971). The two Pleistocene orders are Apterygiformes and Caprimulgiformes, although Brodkorb (1971) considers that all the orders of birds may have been established by the end of the Eocene, 36 million years ago. Open bars are living orders; hatched bars are extinct orders.

Birds have been called 'flying reptiles', for not a single major new structure has appeared in them since they arose from the reptiles, about 150 million years ago (Figure 39). The changes in musculature, central nervous system and sense organs that have occurred are merely avian modifications of the reptilian heritage (Mayr, 1960). On the other hand, they have evolved a constant temperature mechanism which is superior to that in the mammals, for it operates at a higher temperature. In place of intelligence, they have acquired an impressive array of 'wired-in reflexes', instinctive behaviour with respect to nest building, fledgling feeding, evasive tactics, migration etc.

In comparison with mammals, birds apparently produce species far more slowly; most Miocene nonpasserine birds belong to modern genera, and many Pleistocene birds cannot be separated specifically from living ones (Mayr, 1967). Perhaps relevant in this connection, Aquadro and Avise (1982) observed that the genetic distance, as shown by protein electrophoresis, is markedly lower in birds than in other vertebrates; one suggestion for this is that the intrinsic rate of protein change has been much reduced.

Since the Pleistocene epoch, birds have declined both in absolute numbers and in variety of species. According to Brodkorb (1971), the number of species has been reduced by 20% in the last 10 000 years. The chief factors in this decline appear to have been an increase in climatic severity and changes in the mammalian population (Stahl, 1974). Successive glaciations in the northern hemisphere have caused many species of birds to die out, while others have shifted their range southward, some adopting a seasonal migration pattern. With the rise of carnivorous mammals, many of the large terrestrial birds in the southern hemisphere have become extinct and, as the large animals (mastodons, cave bears etc.) have declined, so have the larger birds of prey (condors, vultures and eagles). They were replaced by smaller hawks and owls better adapted to the capture of rodents, which became the more numerous mammals. But even these smaller raptors are coming under threat from man-made agricultural chemicals

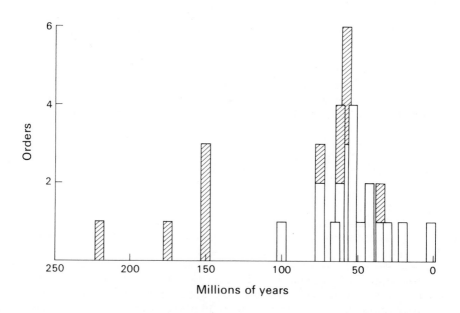

Figure 40 The times of first appearance of orders of mammals, according to the fossil record (data from Romer, 1966 and Olson, 1971) and molecular evidence (data from Goodman et al., 1982). The earliest time has been taken for each order. Open bars are living orders; hatched bars are extinct orders.

which affect calcium metabolism and thus the structural integrity of their eggs. Although the birds may be still further reduced in numbers and variety, they will continue to evolve in minor ways and to maintain their place, because their main aerial habitat is not threatened by any other form of life.

Mammals

Although there are only about 4500 species, the mammals in their great diversity of form and function are perhaps the most adaptively successful group in the animal kingdom. They reached their greatest number of species in the Eocene and Oligocene, but they have been declining ever since (Figure 40). Today only one-third of the known mammalian genera are still living, but the frequent extinctions may partly be a necessary consequence of the continuing evolution of the group. Until recently, the mammals appeared to be still as dominant on the land as they were 50 million years ago, but some writers (Schindewolf, 1950; Müller, 1955) have expressed pessimism regarding their future. Huxley (1943) concluded that most mammalian lines cut themselves off during the Tertiary from much further progress by concentrating on immediate specialisation. On the other hand, Stahl (1974) considers that the now relatively impoverished condition of the mammals is mainly due to the influence of man.

TABLE 15

Doses of X-rays in rem (rontgen equivalent mutagenic) giving 50% lethality in 30 days (abridged from Bacq and Alexander, 1961).

Organism	LD_{50}		
Bacterium *(Escherichia, Bacillus)*	5 600	–	150 000
Protozoan (amoeba, *Paramecium*)	100 000	–	300 000
Fungus (yeast)			30 000
Mollusc (snail)	8 000	–	20 000
Fish (goldfish, *Carassius*)	670	–	1 800
Amphibian (frog, newt)	700	–	3 000
Bird (fowl)	600	–	800
Mammal(pig, rabbit)	250	–	800

Undoubtedly the actions of mankind will constitute the major factor in future mammalian evolution. Since 1600, about 225 species of vertebrates are known to have become extinct and the rate of extinction increased steadily until about 1920, when conservation plans were put into effect (Thornback and Jenkins, 1982). Many of the earlier extinctions were caused by hunting, but an important reason for recent mammalian decline has been the accelerating alteration of habitats, especially the tropical forests, of which 40% have been destroyed in the past 150 years. Future disastrous threats to the mammals are the possibilities of extensive radioactive contamination or even atomic war, because mammals are much more sensitive to radiation than are birds, fishes, amphibia or reptiles (Bacq and Alexander, 1961; Table 15).

Future mammalian evolution is difficult to assess because of the pervasive influence of man, especially on the larger animals. On the one hand, the numbers and habitats of many animals with longer generation times have been so severely reduced that further genetic adaptive change can only be minimal, and they will be fortunate to survive as species. Animal groups that have been reduced to small numbers are said to be prone to extinction due to lack of genetic variation, as is the probable case in the cheetah (O'Brien et al., 1985). On the other hand, there has been a widescale mixing of indigenous with wild or feral animals from very different environments, which may foreshadow an accelerated evolution in some of the smaller, faster-breeding species. Such abrupt competition can also be to the disadvantage of more primitive mammals, such as the marsupials, with relatively antiquated adaptive systems in competition with eutherian forms.

It seems reasonable to suppose that the future of mammalian evolution lies with two groups, rodents and man. Even before *Homo sapiens* appeared, the typical rats and mice became, in the later Cenozoic and Recent times, the most abundant of all mammals, in numbers of genera and species. With the advent of civilised, i.e. town-dwelling, man, certain species of these rodents have become even more numerous. The brown rat in particular, one of 270 species of *Rattus*, with its unspecialised form, omnivorous habits, fierce and cunning behaviour, great fecundity and enormous numbers, seems of all organisms now living the one with the greatest evolutionary potential; its further evolution appears to be assured.

The past evolution of man from advanced nonhuman primates has been rather unremarkable except for an accelerated development of the brain, both quantitatively and qualitatively. The enlargement of the brain in the primates, however, does not seen to be exceptional, because in much the same period (Eocene to Recent) the brains of Artiodactyla, Perissodactyla and Carnivora also evolved steadily to much greater size relative to the body size (Jerison, 1970). What does seem to be novel is a capacity in man for abstractive thinking on remembered information. Otherwise, man and the chimpanzee have much the same reaction times (speed of response to a signal), remembrance of a well-practised task and ability for uncomplicated associations known as classical conditioning (Nissen, 1955). Considering the relatively short period from the nonhuman primate ancestors, the Australopithecinae (about one million years), this novel qualitative change in the brain of the genus Homo must have been due to one or more mutations in cerebral functioning. These mutations are thought to have occurred more than 300 000 years ago, because an analysis of palaeolithic art and hand-axe manufacture suggests that the mental ability of the species of

Homo living then was equivalent to that of modern man (Gamble, 1980). Whatever its origin, this new activity of the brain has now practically suppressed the biological evolution of *Homo sapiens.*

If, as may happen, circumstances allow man to evolve further, the main influences in his morphological evolution are likely to come from enhanced mental development and the degeneration of certain features of structure. According to Hrdlicka (1929), who made predictions based on the past few thousand years of human evolution, there could be a moderate expansion in brain capacity with correlated changes in internal organisation and an improvement in blood supply. This would tend to result in a thinner and larger skull, as would a reduction in the jaws and their muscles due to the eating of better prepared foods. Together with sexual selection, the latter influence could lead to a decrease in the size of some of the bony parts of the face, giving individuals with deeper set eyes, narrower noses, smaller jaws, prominent chins and fewer and more irregular teeth. The hair of the head could further disappear and the body become more slender, with hands and feet becoming narrower and the fifth toe diminishing more in size. Such evolutionary trends at present remain latent.

Instead of the continuation of pre-existing trends, evolution in man in appropriate circumstances may be directed towards physical and mental improvement. Man has defects in both his physical construction and his mental makeup, which a further period of natural selection might rectify. The physical drawbacks, which are the result of overgrowth of the brain, include narrow and twisted nostrils, overcrowding of teeth, and air-filled cavities (sinuses) in the skull that are liable to infection and inflammation or decay. Other handicaps, proneness to rupture, prolapse and varicose veins, and damage to feet, are brought about by the assumption of an upright posture (Haldane, 1947). But what is most damaging to man's prospects for survival is his retention of a number of primitive emotions and impulses, from his anthropoid ancestors. The animal emotions of fear and hatred, to which are added the peculiarly human ones of anger and jealousy, are now threatening his future. According to Goudge (1961), man is a badly adapted creature because his mental evolution has been too rapid for him to divest himself of these primitive responses, which probably served to enhance his early survival.

It is unfortunate for man's future evolution that, although genetic changes still occur in human populations due to differential mortality and fertility, and to interracial breeding, they are probably not of much importance. Genetic adaptation has been replaced by a far quicker and more powerful adaptive process, that of the cultural transmission of remembered information.

Vertebrate molecular genetics
Vertebrates have existed only for the last 400 to 500 million years, but it is apparent that their unique organ systems and tissues have depended on the appearance of new proteins. For example, vertebrate blood plasma contains proteins to protect red blood cells and proteins for the maintenance of the organism (albumin, fibrinogen, transferrin, haptoglobin

TABLE 16

The modular construction of vertebrate proteins of the serine protease family.

	Module or Domain	Structure	Function	Occurrence
A	Signal peptide domain	Hydrophobic polypeptide of 18–23 amino acids	Protein transport out of cell	Bloodclotting factor X Thrombin Plasminogen activators Trypsin Kallikrein Chymotrypsin Elastase Haptoglobin
B	Zymogen peptide domain	Contains the catalytic amino acids, histidine, aspartate and serine	Protein cleavage	Prothrombin Plasminogen Plasminogen activator Bloodclotting factor X Protein C
C	Regulatory domains			
	1 Kringle domain	85-amino acid polypeptide linked by three disulphide bridges	Arginine or lysine binding	Prothrombin Plasminogen Plasminogen activator Bloodclotting factor XII Bloodclotting factor X
	2 Calcium binding domain (vitamin K dependent)	47-amino acid polypeptide with one disulphide bridge	Phospholipid and protein binding	Bloodclotting factor X Bloodclotting factor VII Prothrombin Protein C Protein S
	3 Epidermal growth factor domain	40-amino acid polypeptide linked by three disulphide bridges	Transmembrane receptor binding	Bloodclotting factor X Bloodclotting factor IX Bloodclotting factor VII Bloodclotting factor XII Bloodclotting factor XI Plasminogen activator Protein C Protein S

and immunoglobulins) which are not found in invertebrates or protochordates. Some of the new genes have arisen by duplication with subsequent point mutations to give proteins similar in structure but different in function or expression, e.g. the globin gene family and fibrinonectin. Others have resulted from the joining of protein domains from several different sources to create new proteins with novel functions, e.g. serine proteases.

Most of the known genetic duplicative processes (breakage and reunion, unequal recombination, reverse transcription and integration and differential replication) are effective in initiating multigene families but are rarely responsible for intragenic duplication. The reduplication of a segment of a gene is greatly facilitated by the presence of intervening sequences; e.g. an unequal crossing-over event within the fourth and second introns of two alleles generated a duplication of 1700 base-pairs in the haptoglobin gene (Maeda et al., 1984). Intervening sequences seem to provide 'hot spots' for recombination because their presence and length can increase the length of a vertebrate gene by 10 to 30 times the coding sequence and increase the frequency of intra- or inter-genetic recombination by a factor of 10^6 to 10^8 (Gilbert, 1987). The serine protease family of genes is particularly prone to acquire intragenic duplications, and it shows as well a great range in modular construction and assembly.

The especially clear patterns of domains in the serine protease family are found in digestive enzymes like trypsin, enzymes of the blood clotting and complement cascades and in polypeptide processing enzymes like kallikrein (Table 16). In some other genes, the intervening sequences do not typically fall between domains, perhaps as a result of removal, insertion or movement of intervening sequences. Deletion and insertion on a small scale caused by slipped mispairing is a common cause of intron movement in such genes as those for ß-globin in man, chorion proteins in silkworms and actin in sea urchins (Moore, 1983). Genes like those for flavin-containing enzymes and for fibrinogens that had duplicated before, or at the time of, the origin of vertebrates, show differences in intron position, but genes that have duplicated since that time have maintained the position of their introns except for some intron sliding; no gain of introns has been observed during the evolution of vertebrates (Irwin et al., 1988). The genes of invertebrates generally have fewer introns than vertebrate genes (Table 12, page 113), probably due to a lower frequency of intron insertion. However, there is good evidence that introns have been removed from some invertebrate genes like those for calcium-binding proteins (Wilson et al., 1988).

Besides their evolutionary role, the presence of intervening sequences allows a single messenger RNA to be translated as several partially different proteins, and this fact may account for the selective advantage of split genes. Alternative RNA splicing has, however, only been found so far in invertebrate (*Drosophila*) and vertebrate (birds and mammals) animals (Breitbart et al., 1987).

A gene that has been formed by the assembly of coding sequences (exons) has most of its intervening sequences in the same phase and is often related to other genes containing similar exons with their intervening sequences also in the same phase (Table 17). The sequences of the intervening sequences surrounding the modules comprising the plasma proteases that are considered to have evolved by module exchange obey these rules without exception (Patthy, 1987). These introns lie between the first and second nucleotides of a codon (phase 1 introns). They are characteristic of the growth factor, calcium-binding and kringle modules and form the most frequent class. Genes other than those for plasma proteases that are believed to have been assembled from such modules include those encoding all members of the immunoglobulin supergene family, epidermal growth factor precursor, low density lipoprotein receptor

TABLE 17
The translational phases of intervening sequences in some
vertebrate genes for which there is evidence of origin by the
duplication and recruitment of coding modules.

	Gene	Intervening Sequences	
		Number	Phases
Phase 0			
	Collagen, helical segment	7	0,0,0,0,0,0,0
	ß-casein	8	0,0,0,0,0,0
Phase 1			
	Prothrombin	13	1,0,1,1,2,1,1,1,2,2,2,1,0
	Low density lipoprotein receptor	17	1,1,1,1,1,1,1,1,2,2,1, 0,1,1,1,0
	Coagulation inhibitor, protein C	8	1,0,1,1,1,0,1
Phase 2			
	Glucagon	5	2,2,2,2

and ß2-glycoprotein I. Genes that have introns lying between two codons
(phase 0 introns) are those for type III collagen, ß-casein and the precursor
of growth hormone. Introns of the last phase type, which lie between the
second and third nucleotides of a codon (phase 2 introns), are found in the
gene for preproglucagon.

It appears that nearly all examples of gene origin in relatively recent
times by the acquisition of sequences represented in other genes are from
the vertebrates, particularly the mammals. Irwin et al. (1988) suggested that
the vertebrates may owe their origin to the evolution of a great number of
novel proteins as the result of the presence or insertion of intervening
sequences in many genes.

4.6 *Conclusion*

This survey of organisms allows rough estimates to be made as to which
groups are static and degenerating, which are static but maintaining their
place, which have the potentiality for further evolution and which are still
evolving. These assessments can only be very approximate, particularly
because the nature and origin of adaptively significant variation is still

unclear and because the actual tempo of evolutionary change is a matter of current debate. Are long periods of stasis succeeded by sudden bursts of adaptive radiation according to the punctuation model, or is evolution a continuously occurring process according to the gradualist view? Even if evolution follows the latter course, it is still discontinuous in a sense because it is evidently not applicable to those contemporary groups of organisms that have remained the same for long periods of geological time. But if evolution only occurs periodically, with the duration of the intervening periods being unknown, there is always the possibility that a static or even a senescent group can start its evolution afresh, even though examples are very rare. This argument is independent of minor or quantitative changes in characters which, although sometimes included under the heading of evolution, are more properly ecotypic adaptations.

4.6.1 *The present status of evolution*
It is evident that most groups or organisms have been evolutionarily static for a very long time. Even the more recently evolved ones such as birds and mammals (except man and rodents) have been in decline for millions of years. So, generally speaking, the impressions of Decugis (1941), Huxley (1943) and Grassé (1973) that evolution has run down, appear to be confirmed. On the other hand, quantitative adaptive changes are still occurring in a wide range of organisms such as mammals, fishes, snakes, plants, and perhaps molluscs and frogs. These changes may merely represent genetic adjustments to slightly changing environments and may not be symptomatic of major evolutionary advance. For, with the exception of the brain of man, there has been nothing really new in evolution for tens of millions of years, i.e. no new metabolic syntheses, new structural plans or new developmental strategies. The inference is that most groups of organisms are now evolutionarily static or in evolutionary equilibrium with their environments. This is, in fact, the expected state that each group will sooner or later reach because it is unlikely that either the evolution or expansion of a group can continue indefinitely. The only argument that has been advanced for an overall continuance of evolution is that of the constantly deteriorating environment. But if the organism is well adapted and if mechanisms have developed to cope with minor environmental changes, there is no necessity for further evolution except in cases of catastrophe, where it is usually impossible anyway. The role of evolution has been, and is, to produce well adapted organisms which fit the chosen environments, dealing effectively with enemies and competitors, and reproducing in a balanced fashion. Such organisms have indeed been produced. A clear example of such a group is the bacteria, which have evolved systems for the horizontal transfer of genes on plasmids and transposons, a mechanism which effectively copes with most changes in the surrounding medium and diminishes the need for mutational change. Consequently, the bacteria, having completed their evolution a long time ago, now remain relatively quiescent in the evolutionary sense. Among the eukaryotes, groups which appear to be in equilibrium by virtue of the fact that they neither evolve nor decline, are some algae, some protozoa, fungi,

many of the invertebrates, lizards, birds and above all, by reason of his cultural evolution, man.

Some organisms may be only temporarily arrested in their further evolution until a way is found around a primarily limiting feature of their construction or physiology. Thus, the advance of the plants was confined to such forms as liverworts, mosses and ferns for about 100 million years, until an independence of water in the sexual process was reached. Other examples of a successful, or at least a compromise solution to an inherent limitation, are the development of communities in insects as a compensation for small individual size and of instinctive behaviour in birds in place of a larger and better brain. Further evolution-blocking characters of this kind still awaiting a solution may include the gelatinous sheaths of *Volvox* and the cleiodoic eggs of birds.

Finally, there are those groups of organisms which are still evolving, or at least have been evolving recently. These are the relatively late-appearing organisms which dominate the earth in land, air and water; they are the mammals, birds and fishes. However, in spite of their dominance and success in terms of numbers of species and individuals, only the fishes have been maintaining their place lately.

4.6.2 *Causes of decline in the progress of evolution*

As already mentioned, some groups of organisms are evolutionarily static but very successful because they are well adjusted to their environment. Additionally, or consequently, they may have evolved systems which minimise a requirement for further evolution or which replace the adaptive role of the evolutionary process. Such systems, of very varying effectiveness, are quite widespread; for example, plasmids and transposons in bacteria, cryptobiosis in nematodes, tardigrades and rotifers, phenotypic flexibility in plants, behavioural patterns in animals and cultural processes in man. Where environmental changes would normally require a genetic or evolutionary response, these systems allow the organism to avoid them (cryptobiosis), absorb them (phenotypic flexibility) or compensate for them (behavioural shifts) or make temporary additions to the genome (bacterial plasmids) to circumvent them.

In addition to the presence of evolution-minimising mechanisms, there are other reasons for the slowing and stopping of evolution which are generally familiar and expected. Perhaps first among them is specialisation, which is clearly a major drawback to the continued evolution of many groups. Although specialisation is usually associated with evolutionary arrest, it may not be disadvantageous to the organism. Many groups rely on specialisation to reach parity with their environments, and it is their specialisations which remove the need for further evolution. Specialisation, therefore, need not lead to extinction, although those groups that have become overspecialised in certain features which inhibit further adaptation have generally become extinct or reduced to relict species. A further cause for the slowing down of evolution has been the filling of most of the earth's adaptive niches. This means that evolution now, and for some considerable time in the past, could only proceed by

TABLE 18

Numbers and phases of intervening sequences in some genes unique to dicotyledonous plants or mammals, respectively.

Gene	Plant	Intervening Sequences Number	Intervening Sequences Phases
Nodulin	Soybean	4	1,1,1,1
Aux 28	Soybean	4	2,1,2,1
Aux 22	Soybean	2	1,2
Phaseolin storage protein	Phaseolus	5	1,0,0,0,1
Extensin	Carrot	1	in noncoding region
Chalcone synthase	Antirrhinum	2	1,0
Gene	Mammal	Number	Phases
Plasminogen activator inhibitor	Man	8	1,1,1,2,1,1,1
Low density lipoprotein receptor	Man	17	1,1,1,1,1,1,1,1,2,2,1, 0,1,1,1,0
Blood coagulation inhibitor, G	Man	8	1,0,1,1,1,0,1
Major histocompatibility complex Aß2	Mouse	4	1,1,1,1
Myosin heavy chain, rod region	Rat	8	1,0,0,0,0,0,0,0
T cell receptor ß chain	Man	3	1,0,0
ß2-microglobulin	Mouse	3	1,1

displacing an established group by a new, more efficient version, a procedure expected to be much more difficult than primary colonisation.

While some static groups such as pseudopodial protozoans and echinoderms are quite specialised in several features, so that their stasis can reasonably be attributed to overspecialisation, others such as green algae, sponges and nematodes are not, nor is their evolution held back by any discernible obstacle. Simpson (1964) attributes a lack of evolutionary change over very long periods to environmental conditions so stable as not to elicit any evolutionary movement. But it is now beginning to appear that there may be internal factors partly responsible for evolution in addition to the external (environmental) ones. They comprise such mechanisms as gene conversion, reverse transcription of RNA, subgenic assembly based on the presence of intervening sequences and transposition, as discussed in Part 3.

In view of the demonstrable ineffectiveness of point mutation, it is proposed that novel evolution requires the formation of new genes by processes of module exchange and assembly brought about by recombination within intervening sequences. Accordingly, the question of evolutionary potential is thought to be directly related to the capacity of the genes of any group to be reassembled to make new genes. The structure of contemporary genes shows that most, if not all, phyla began their evolution with the construction of new genes by the exchange and assembly of coding modules. However, such assembly depends upon the number, size and translational phase of the intervening sequences that the genes involved possess. In most kingdoms or phyla one or more of these properties of recombining intervening sequences have gradually decreased. The first-formed genes unique to bacteria and many protista appear to have been constructed by such 'exon shuffling', but, with a loss of intervening sequences, this process stopped and evolution slowed down. Among those groups where intervening sequences remain frequent, the filamentous fungi, flowering plants and vertebrates, the chances of new assemblies by recombination may be greatly decreased by a reduction in size; e.g. the introns of filamentous fungi are usually 48 to 240 base-pairs, those of *Drosophila* often less than 100 base-pairs, while vertebrate introns, although very variable in size, can be up to 10 000 to 20 000 base-pairs in length. To the extent that novel evolution is due to the origin of new genes and that subgenic assembly is a major mode of such origin, the evolutionary potential of flowering plants and mammals would be expected to be much higher than that of contemporary protozoa and invertebrates, as indeed seems to be the case. However, genes that appear to have been uniquely evolved in vertebrates generally have a greater number of introns and more uniformity of intron phase that do those evolved in flowering plants (Table 18).

GENERAL CONCLUSIONS

The present ineffectual state of point mutation for evolution can be attributed to the fact that there have been successive changes in the targets and mechanisms of evolution. Initially, there was no alternative to point mutation, with a frequency very much higher than that of today, to make the small prototype genes encoding primitive metabolic enzymes. As the enzymes have become larger in order to acquire increased specificity, regulatory and co-enzyme sites, and membrane-binding regions, nucleotide replacement has become more damaging to the enzymes and has been much decreased in frequency. Even the RNA-amino acid code has probably altered to become more absorptive of the destructive consequences of nucleotide substitution. Another factor increasing the damaging effects of point mutation on the survival and reproduction of the individual was the steady increase in the number of genes in the genome. It has been calculated that about 500 genes would be required to specify the number of reactions needed for life on a preformed source of carbon, but between 50 000 and 100 000 genes are required to construct and maintain a mouse.

With a decline in the efficacy of nucleotide substitution, other mechanisms for providing variation were required to maintain or enhance rates of evolution. The action of some of these mechanisms, and their nature, can be deduced from the structures and affinities of contemporary genes and proteins. Examples are such protein domains as the haem-binding structure of oxygen- and electron-carriers like myoglobin, haemoglobin and cytochromes, or the mononucleotide co-enzyme fold of carbohydrate-metabolising enzymes. These domains can be traced in anciently evolved proteins before the plant and animal kingdoms diverged, because the three-dimensional structure of co-enzyme or substrate-binding regions is conserved to a much greater extent than is the amino acid sequence. Although more than 2000 different enzymes are known, they are variants of a few families of related molecules, the members of which, characterised by the same bond-rearranging mechanisms but with differences in substrate specificity, have been made by a process of gene duplication. Related specificities can be brought about by nucleotide substitutions following gene duplication, but entirely new specificities appear to need a novel assembly of nucleotide sequences determining particular protein domains. For example, many vertebrate proteins concerned with blood clotting and the immune system, do not occur in invertebrates or protochordates. The nature of their sequences and structures is consistent with the corresponding genes being formed by modular processes acting on duplicated genes about 500 million years ago.

The modular system is also used in evolving regulatory genes. In bacteria and their viruses, a common module is responsible for the inversion or deletion of DNA segments to give flagella variation or a change in bacteriophage host range. Modules of information are used in assembling bacterial genes whose products respond to environmental stimuli, such as change in osmolarity of the medium, nitrogen or phosphate limitation and the presence of toxic compounds. The signal received by the receptor portions of the regulatory genes is then transmitted to a second regulatory region which probably interacts with RNA polymerase and binds to the DNA to activate appropriate gene transcription. These proteins have apparently evolved by recruiting a sensor module specifying about 200 amino acids and a regulatory module for 120 amino acids from various progenitor genes; the remainder of the proteins vary in length and are not homologous. In invertebrates, the Notch gene of Drosophila, which regulates the differentiation of epidermal-like tissues, contains a sequence homologous with the growth hormone-coding portion of the epidermal growth factor genes. In vertebrates also, some genes or genomes have had inserted in them one or more units of genetic information for a specific function. For example, in mammals, epidermal growth factor, enzyme proteases involved in blood clotting and fibrinolysis, and vaccinia virus all contain a closely related sequence which may specify transmembrane binding.

While point mutation must have been effective in primordial times, nowadays, except for the fine adjustment of proteins, it has been replaced by various sorts of recombinational mechanisms for generating genetic variation. Moreover, different mechanisms appear to be involved in the formation of different taxons, at least in animals.

The production of varieties resistant to toxins developed by man takes less than a few tens of years, and is generally by point mutation that inactivates the gene that makes the protein target or that decreases the affinity of the protein for the toxin. The breeding of animals over the last few thousand years has resulted in changes which also may be caused by point mutations in trans-acting proteins, particularly hormones. When genetic differences between species are examined, the majority seem to be in cis-disposed binding sequences which specify the timing and place of gene expression. Because of the usual multiplicity of such sequences, changes in them cannot be due to point mutations, although nucleotide substitution added to a process like conversion could be effective. Genetic differences characteristic of animal orders affect cell lineages in invertebrates, cellular and tissue interactions in vertebrates, and the chronology of development. They appear to be due to new or altered regulatory processes which, because they must fit with existing patterns of gene expression, may be caused by the modular assembly or exchange of existing informational sequences.

The above examples indicate that as time goes on, newer and rarer mechanisms with greater phenotypic expression sometimes come into action, depending on the group and its evolutionary stage. In all organisms, evidence of the times or origin of classes, and in most organisms of the times of origin of orders, both of which are characterised by one or more qualitatively new features, shows that innovative evolution usually slows down or stops after undergoing an early surge. It appears that when organisms are brought into equilibrium with their environments, they tend to resist further evolution as the environment changes. The advantage of such stasis can be understood when it is considered that the evolution of even a minor modification, that is, the selective incorporation of the modification into all members of the population, incurs a large biological cost. This cost is expressed as a number of genetic deaths or their equivalents, estimated to be equal to 10 to 20 times the number of individuals living at any one time.

The view of at least some neoDarwinists is that, because of the close adaptation of organisms to their environments, practically any change in the environment's physical or biological nature will be deleterious. Such changes occur continually, owing to the constant rate of production of disadvantageous mutations, sporadic alterations in geology or climate, or the adaptive evolution of other species. It is proposed, therefore, that the environment of a species is continually deteriorating, and all species need to evolve continually in order to maintain their adaptive position. However, a formal evolutionary response can be avoided if an alternative process like plasmid gene transfer in bacteria or cultural adaptation in man is available. In other nonevolving groups or organisms, evolution-dampening devices such as cryptobiosis, phenotypic flexibility and homeostasis occur, but the real processes limiting evolution in some of the algae, protozoa, fungi and invertebrates are not known. Developmental constraints and the difficulty of shifting from colonising to replacement evolution have been suggested as causes of stasis.

A possible alternative explanation is provided by the capacity to form new genes in different groups depending on the ability of the group to utilise mechanisms of subgenic assembly. The evidence advanced in the latter part of this book suggests that this process may depend upon the number, size and translational phase of existing intervening sequences. The validity or otherwise of this suggestion may become apparent when there is a more complete knowledge of the structure and mode of formation of genes in contemporary organisms.

GLOSSARY

Alkylation The introduction into an organic compound of an organic group derived by the removal of a hydrogen atom.

Allele or Allelomorph One of two or several forms of a gene at a particular position (locus) on a given chromosome.

Allosteric site A region on some enzymes that is distinct from the active site and to which small molecules may bind to activate or inhibit the enzyme.

Amino-terminal The end of a polypeptide chain that contains a free α-amino group.

Anabolism Enzymatic processes involved in the synthesis of compounds.

Ångström A unit of submicroscopic distance equal to 10^{-10} metres.

Bacteriophage or Phage A virus that is an obligate intracellular parasite of a bacterium.

Blastula The early animal embryo preceding the gastrula and consisting of a hollow ball of cells with a wall several cells thick.

Carboxy-terminal The end of a polypeptide chain that contains a free α-carboxyl group.

Catabolism Enzymatic processes involved in the breakdown of compounds to provide energy.

Chromatin Fibres of nucleic acid and protein comprising eukaryotic chromosomes.

Cis- Indicating the position of a DNA sequence involved in regulation, in relation to the gene whose expression is affected, i.e. a *cis*-disposed sequence is on the same DNA molecule as the gene itself and is usually a protein-binding site.

Codon redundancy or degeneracy Two or more nucleotide triplets (codons) that code for the same amino acid.

Constitutive enzyme An enzyme that is synthesised continuously regardless of the presence of its substrate.

Conversion The replacement during replication of a particular DNA sequence within a gene by a sequence from an allelic or sequence-related gene.

Dimer A compound resulting from the association of two identical simpler compounds (monomers).

DNA repair The excision or replacement of one or more nucleotides from a DNA molecule of incorrect or damaged sequence.

Endonuclease An enzyme that catalyses the breakdown of nucleic acid by making internal cuts.

Eukaryote An organism that contains its genetic material within a nucleus, has nuclear membranes and membrane-bound organelles (mitochondria and, if a plant, chloroplasts).

Exon A region of DNA in a gene that is translated into protein and is separated from other regions by nontranslatable sequences.

Exonuclease An enzyme that catalyses the breakdown of nucleic acid from ends of the strands.

Gastrula The early animal embryo succeeding the blastula in which cells are differentiated into the germ layers and the central cavity.

Genome The haploid set of chromosomes and genes.

Genotype The genetic constitution of an organism.

Haploid A cell or organism containing a single set of chromosomes.

Hydrophilic Designates molecules or groups that tend to associate with water.

Hydrophobic Designates molecules or groups that are poorly soluble in water.

Inducible enzyme An enzyme that is synthesised or has its synthesis increased by the presence of an inducing agent such as its substrate.

Intervening sequence or Intron A region of DNA in a gene that is excised after transcription and that does not code for amino acids in a protein.

Intron See Intervening sequence.

Mutator gene A gene that increases the rate of mutation in other genes.

Nematocyst The stinging part of a specialised cell found in the Cnidaria (hydroids, jellyfish, corals etc.).

Neoteny The persistence in an adult organism of juvenile or larval features.

Nucleophilic Having an affinity for positive electric charge.

Operon A group of closely linked genes producing a single messenger RNA when induced.

Phenotype The sum of the morphological or physiological expressions of the genes of an organism in interaction with the environment.

Phosphodiester bond The bond by which nucleotides are linked together through a bridging phosphate group.

Plasmid A self-replicating genetic element of bacterial cells.

Pleopod A limb appendage, usually for swimming or respiratory function, on the abdomen of the Crustacea.

Polyadenylation The addition of adenine nucleotides to the 3' end of messenger RNA in eukaryotes.

Polycistronic Designates the carrying of information of two or more genes on a single messenger RNA.

Polymerase DNA polymerases are enzymes that catalyse the formation of phosphodiester bonds in the synthesis of DNA. RNA polymerases catalyse the synthesis of RNA, usually from DNA templates.

Polynucleotide A molecule consisting of two or more nucleotides joined by phosphodiester bonds.

Polypeptide A molecule consisting of two or more amino acids joined by covalent bonds between the amino group of one and the carboxyl group of the next.

Position effect An alteration in gene expression due to a change in position of a gene on a chromosome, as by translocation or inversion.

Prokaryote Generally unicellular organisms that comprise various bacteria and cyanobacteria and that lack nuclear membranes and organelles.

Proton An elementary particle with a positive charge that occurs in the nuclei of atoms.

Recessive gene An allele whose expression is masked by the presence of another allele at the locus.

Repressible enzyme An enzyme that has its synthesis blocked or decreased by the presence in the cell of certain compounds.

Repressor The protein product of a gene that can combine with an inducing substance or with a nucleic acid sequence (operator) adjacent to, or part of, a gene.

Tautomerisation The reversible conversion of one molecule into another of the same molecular formula by an alteration in the position of a proton or other elementary particle.

Trans– Indicating the position of a DNA sequence involved in regulation, in relation to the gene whose expression is affected, i.e. a *trans*-disposed sequence is on a different DNA molecule from the gene it affects and usually codes for a protein.

Transcription The process whereby genetic information in DNA is converted into a complementary single strand of RNA.

Translation The process whereby genetic information in RNA is converted into a linear sequence of amino acids to form a protein.

Transposon A bacterial genetic element, usually carried on a plasmid, that can move to other parts of the genome.

The Genetic Code for RNA

The genetic code for RNA and the amino acids specified by the triplets are shown below. Except for proline, where the full structure is drawn, only the amino acid side-chains are shown.

	U	C	A	G	
U	$-CH_2-$⬡ 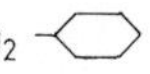 Phe ---- $-CH_2-CH{<}^{CH_3}_{CH_3}$ Leu	$-CH_2-OH$ Ser	$-CH_2-$⬡$-OH$ Tyr ---- Ochre ---- Amber 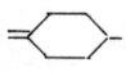	$-CH_2-SH$ Cys ---- Opal ---- $-CH_2-$indole Trp 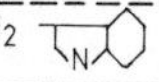	U C A G
C	$-CH_2-CH{<}^{CH_3}_{CH_3}$ Leu	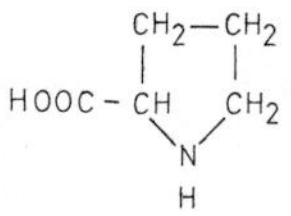 $HOOC-CH{<}^{CH_2-CH_2}_{N(H)}CH_2$ Pro	$-CH_2-$imidazole His ---- $-(CH_2)_2-CONH_2$ Glu 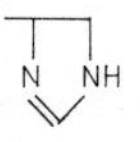	$-(CH_2)_3-NH-C{<}^{NH}_{NH_2}$ Arg	U C A G
A	$-CH{<}^{CH_2-CH_3}_{CH_3}$ Ile ---- $-(CH_2)_2-S-CH_3$ Met	$-CH{<}^{CH_3}_{OH}$ Thr	$-CH_2-CONH_2$ Asn ---- $-(CH_2)_4-NH_2$ Lys	$-CH_2-OH$ Ser ---- $-(CH_2)-NH-C{<}^{NH}_{NH_2}$ Arg	U C A G
G	$-CH{<}^{CH_3}_{CH_3}$ Val	$-CH_3$ Ala	$-CH_2-COOH$ Asp ---- $-(CH_2)_2-COOH$ Gln	$-H$ Gly	U C A G

APPENDIX 2
The Geological Time Scale

The chronological ages indicate the beginning of an epoch, period, era or eon, and are those given by Palmer (1983). The Ediacarian period is from Cloud and Glaessner (1982) and the Proterozoic eon from Awramik (1981).

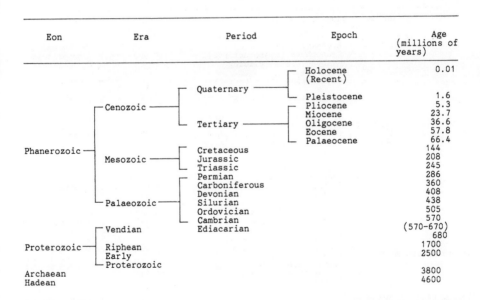

Eon	Era	Period	Epoch	Age (millions of years)
		Quaternary	Holocene (Recent)	0.01
	Cenozoic		Pleistocene	1.6
		Tertiary	Pliocene	5.3
			Miocene	23.7
			Oligocene	36.6
			Eocene	57.8
			Palaeocene	66.4
Phanerozoic	Mesozoic	Cretaceous		144
		Jurassic		208
		Triassic		245
		Permian		286
	Palaeozoic	Carboniferous		360
		Devonian		408
		Silurian		438
		Ordovician		505
		Cambrian		570
	Vendian	Ediacarian		(570–670)
				680
Proterozoic	Riphean			1700
	Early Proterozoic			2500
Archaean				3800
Hadean				4600

177

APPENDIX 3
The Kingdoms and Phyla of Organisms

The kingdoms and phyla of organisms mainly according to Whittaker (1969), together with estimates of the number of living species (Altman and Dittmer, 1972) and their time of origin on fossil evidence (L = Lower, M = Middle, U = Upper).

Kingdom, Phylum or Division		Number of Species	Time of origin
Kingdom Monera, Division	Archaebacteriae	?	L. Proterozoic
	Cyanophyta (cyanobacteria)	17 000	L. Proterozoic
	Myxobacteriae (gliding bacteria)	?	?
	Eubacteriae (true bacteria)	2 700	M. Archaean
	Actinomycota (mycelial bacteria)	?	Ediacarian
Kingdom Protista, Phylum	Spirochaetae (spirochetes)	?	?
	Euglenophyta (euglenoids)	370	M. Proterozoic
	Chrysophyta (golden algae)	5 700	Permian
	Pyrrophyta (dinoflagellates etc.)	1 000	Ordovician
	Sporozoa (sporozoans)	3 600	?
	Cnidosporida (cnidosporidians)	1 100	?
	Sarcodina (rhizopods)	17 650	L. Ordovician
	Rhodophyta (red algae)	2 900	M. Proterozoic
	Phaeophyta (brown algae)	1 100	M. Proterozoic
	Chlorophyta (green algae)	5 800	Ordovician
	Myxomycetes (slime moulds)	450	L. Carboniferous
	Phycomycetes	1 250	Silurian
	Charophyta (stoneworts)	207	U. Silurian
	Ciliophora (ciliates etc)	6 000	U. Ordovician
Kingdom Fungi, Phylum	Ascomycetes (sac fungi)	13 500	L. Carboniferous
	Basidiomycetes (club fungi)	15 000	U. Carboniferous
	Deuteromycetes (fungi imperfecti)	10 200	U. Carboniferous
Kingdom Plantae, Division	Bryophyta (liverworts and mosses)	22 000	L. Carboniferous
	Psilophyta (psilophytes)	3	L. Silurian
	Lycopodiophyta (club mosses)	1 070	U. Devonian
	Equisetophyta (horse tails)	20	L. Carboniferous
	Pteridophyta (ferns)	9 000	Carboniferous

Appendix 3 continued

	Class or order	Number of species	Time of origin
	Gymnospermae (pines etc.)	700	U. Triassic
	Angiospermae (flowering plants)	216 000	Triassic
Kingdom Animalia, Phylum	Mesozoa	50	?
	Porifera (sponges)	6 300	M. Cambrian
	Cnidaria (coelenterates)	8 000	M. Vendian
	Ctenophora (comb jellies)	85	?
	Platyhelminthes (flatworms)	13 800	U. Vendian
	Nemertea or Rhynchocoela (ribbon worms)	700	U. Carboniferous
	Acanthocephala (spiny-headed worms)	7 000	?
	Entoprocta (pseudocoelomate polyzoans)	67	?
	Entoprocta or Bryozoa (coelomate polyzoans)	3 750	L. Ordovician
	Brachiopoda (lamp shells)	245	L. Cambrian
	Phoronida (horseshoe worms)	16	Ordovician
	Mollusca (clams, snails etc)	75,700	L. Cambrian
	Sipunculoidea (peanut worms)	262	M. Devonian
	Echiuroidea (spoon worms)	97	U. Vendian
	Annelida (segmented worms)	7 000	M. Cambrian
	Pogonophora (deep-sea worms)	61	U. Vendian
	Chaetognatha (arrow worms)	50	U. Carboniferous
	Arthropoda (insects, crabs etc.)	842 000	L. Cambrian
	Echinodermata (starfish, sea urchins etc.)	5 700	L. Cambrian
	Hemichordata (acorn worms)	85	M. Cambrian
	Chordata Subphylum Tunicata (sea squirts)	1 400	U. Cambrian
	Subphylum Cephalochordata	13	?
	Subphylum Vertebrata	41 400	U. Cambrian

APPENDIX 4
The Classes of Vertebrates and Orders of Mammals

The classes of the Vertebrata according to Romer (1966) and the orders of the class Mammalia according to Lawlor (1979), together with estimates of the number of species and times of origin of each.

		Class or order	Number of Species	Time of origin
Class		Agnatha (jawless fishes)	50	U. Cambrian
		Chondrichthyes (cartilaginous fishes)	550	M. Devonian
		Osteichthyes (bony fishes)	20 000	L. Devonian
			1	L. Devonian
			6	L. Devonian
		Amphibia (frogs, salamanders etc.)	2 300	L. Carboniferous
		Reptilia (snakes, lizards etc.)	5 800	L. Carboniferous
		Aves (birds)	8 600	U. Jurassic
Mammalia			4 090	M. Jurassic
	Order	Monotremata (platypus, echidna)	6	Mesozoic
		Marsupalia (kangaroos etc.)	240	M. Cretaceous
		Edentata (anteaters, sloths etc.)	31	Palaeocene
		Pholidota (pangolin)	8	L. Palaeocene
		Lagomorpha (rabbits etc.)	64	L. Eocene
		Rodentia (rats etc.)	1 600	L. Eocene
		Primates (apes, man)	119	L. Palaeocene
		Dermoptera (flying lemurs)	2	L. Palaeocene
		Chiroptera (bats)	850	M. Palaeocene
		Insectivora (hedgehog, shrew etc.)	400	U. Cretaceous
		Carnivora (cats, seals etc.)	279	M. Palaeocene
		Cetacea (whales, dolphins etc.)	70	L. Eocene
		Tubulidentata (aardvark)	1	M. Miocene
		Artiodactyla (camel, sheep etc.)	172	L. Eocene
		Perissodactyla (horse etc.)	16	M. Eocene
		Hyracoidea (hyraxes or conies)	11	M. Eocene
		Proboscidea (elephant)	2	L. Eocene
		Sirenia (manatees and dugongs)	5	L. Eocene

REFERENCES

Abplanalp, H. (1974) Inbreeding as a tool for poultry improvement. In *First World Congress on Genetics applied to Livestock Production* pp. 897-908. Graficas Orbe, Madrid.

Alberch, P. (1986) The evolution of amphibian metamorphosis: opportunity within constraint. In *Development as an Evolutionary Process*. ed. R.A. Raff and E. Raff, MBL Lectures in Biology Series.

Altman, L. and Dittmer, D.S., Editors (1972). *Biological Data Book*, Vol. 1, 2nd Ed., Fed. Am. Soc. Exp. Biol. Bethesda, Maryland.

Ambros, V. and Horvitz, H.R. (1984) Heterochronic mutants of the nematode *Caenorhabditis elegans*. *Science (Wash.)* 226, 409-416.

Amrhein, N., Johänning, D., Schab, J. and Schulz, A. (1983) Biochemical basis for glyphosate-tolerance in a bacterium and a plant tissue culture. *FEBS Lett.* 157, 191-196.

Ananthan, J., Goldberg, A.L. and Voellmy, R. (1986). Abnormal proteins serve as eukaryotic signals and trigger the activation of heat shock genes. *Science (Wash.)* 232, 522-524.

Aquadro, C.F. and Avise, J.C. (1982) Evolutionary genetics of birds VI. A reexamination of protein divergence using varied electrophoretic conditions. *Evolution*, 36, 1003-1019.

Armaleo, D. (1987) Structure and evolution of prokaryotic and eukaryotic RNA polymerases: a model. *J. Theor. Biol.* 127, 301-314.

Austin, R.B., Morgan, C.L., Ford, M.A. and Bhagwat, S.G. (1982) Flag leaf photosynthesis of *Triticum aestivum* and related diploid and tetraploid species. *Ann. Bot. (London)* 49, 177-189.

Avise, J.C. and Ayala, F.J., (1975) Genetic change and rates of cladogenesis. *Genetics* 81, 757-773.

Awramik, S.M. (1981) The origins and early evolution of life. In *The Cambridge Encyclopedia of Earth Sciences*, ed. D.G. Smith pp. 349-362. Cambridge University Press, Cambridge.

181

Ayala, F.J. (1984) Molecular polymorphism: How much is there and why is there so much? *Dev. Genet.* <u>4</u>, 379-391.

Bacq, Z.M. and Alexander, P. (1961) *Foundations of Radiobiology.* Pergamon Press, London.

Baker, M.E. (1988) Similarity between the estrogen receptor and the DNA-binding domain of the tetracycline repressor. *Biochem. Biophys Res. Comm.* <u>150</u>, 463-468.

Baldwin, E. (1949) *An Introduction to Comparative Biochemistry.* Cambridge University Press, Cambridge.

Baltimore, D. (1985) Retroviruses and retrotransposons: the role of reverse transcriptase in shaping the genome. *Cell* <u>40</u>, 481-482.

Barahona, I., Soares, H., Cyrne, L., Penque, D., Denoulet, P. and Rodrigues-Pousada, C. (1988) Sequence of one α- and two β-tubulin genes of *Tetrahymena pyriformis*. *J. Mol. Biol.* <u>202</u>, 365-382.

Bardwell, J.C.A. and Craig, E.A. (1984) Major heat shock gene of *Drosophila* and the *Escherichia coli* inducible DNA K gene are homologous. *Proc. Nat. Acad. Sci. USA* <u>81</u>, 845-852.

Baroin, A., Perasso, R., Qu, L.-H., Burgerolle, G., Bechellerie, J.-P. and Adoutte, A. (1988) Partial phylogeny of the unicellular eukaryotes based on rapid sequencing of a portion of 28S ribosomal RNA. *Proc. Nat. Acad. Sci. USA* <u>85</u>, 3474-3478.

Barrington, E.J.V. (1964) *Hormones and Evolution.* English University Press, London.

Barrington, E.J.W., (1967) *Invertebrate Structure and Function.* Nelson, London.

Barstad, J.A.B. (1978) Pesticides and heavy metals as food contaminants. In *Toxicological Aspects of Food Safety*, ed. E.J. Leonard. *Arch. Toxicol.* suppl. 1, pp 47-54. Springer, Berlin.

Batschelet, E., Domingo, E. and Weissmann, C. (1976) The proportion of revertant and mutant phage in a growing population, as a function of mutation and growth rate. *Gene* <u>1</u>, 27-32.

Baulcombe, D.C. Martienssen, R.A. Huttly, A.M., Barker, R.A. and Lazarus, C.M. (1986) Hormonal and developmental control of gene expression in wheat. *Philos. Trans R. Soc. London* <u>B314</u>, 441-451.

Baum, J.A. and Giles, N.J. (1985) Genetic control of chromatin structure 5' to the qa-x and qa-2 genes of *Neurospora*. *J. Mol. Biol.*<u>182</u>, 79-89.

Belyaev, D.K. (1980) Destabilizing selection as a factor of domestication. *Proc. XIV Int. Congr. Genet.* <u>1</u>, 64-80.

Bender, W. (1985) Homeotic gene products as growth factors. *Cell* <u>43</u>. 559-560.

Benne, R., Van Den Burg, J., Brakenhoff, J.P.J., Sloof, P., Van Boom, J.H. and Tromp, M.C. (1986) Major transcript of the frameshifted cox II gene from Trypanosoma mitochondria contains four nucleotides that are not encoded in the DNA. *Cell* <u>35</u>, 819-826.

Benveniste, R. and Davies, J. (1973) Aminoglycoside antibiotic-inactivating enzymes in actinomycetes similar to those present in clinical isolates of antibiotic-resistant bacteria. *Proc. Nat. Acad. Sci. USA.* 20, 2276-2278.

Berg, R.R. and Walters, L.E. (1983) The meat animal: changes and challenges. *J. Anim. Sci.* 57 (suppl. 2), 132-146.

Bernardi, F. and Ninio, J. (1978) The accuracy of DNA replication. *Biochimie* 60, 1083-1095.

Bernstein, S.I., Hansen, C.J., Becker, K.D., Wassenberg, D.R., Roche, E.S., Donady, J.J. and Emerson, C.P. (1986) Alternative RNA splicing generates transcripts encoding a thorax-specific isoform of *Drosophila melanogaster* myosin heavy chain. *Mol. Cell. Biol.* 6, 2511-2519.

Beinz, M. and Pelham, H.R.B. (1982) Expression of the *Drosophila* heat shock protein in *Xenopus* oocytes: conserved and divergent regulatory signals. *EMBO J.* 1, 1583-1588.

Blanco, C., Ritzenthaler, P. and Mata-Gilsinger, M. (1985) Nucleotide sequence of a regulatory region of the uid A gene in *Escherichia coli* K12. *Mol. Gen. Genet.* 199, 101-105.

Bloch, R. (1965) Histological foundations of differentiation and development in plants. *Handb. Pflanzenphysiol.* 15/1, 146-188.

Blum, H.F. (1955) *Time's Arrow and Evolution.* Princeton University Press, New Jersey.

Boedtker, H. and Aho, S. (1984) Collagen gene structure: the paradox may be resolved. *Biochem. Soc. Symp.* 49, 67-84.

Bonner, J.T. (1958) *The Evolution of Development.* Cambridge University Press, Cambridge.

Bonner, J.T. (1965) *The Molecular Biology of Development,* Clarendon Press, Oxford.

Bonner, J.T. (1968) Size change in development and evolution. *Paleontol. Soc. Mem.* No. 2, pp. 1-15.

Bonner, J.T. (1974) *On Development.* Harvard University Press, Massachusetts.

Bookstein, F.L., Gingerich, P.D. and Kluge, A.G. (1978) Hierarchial linear modelling of the tempo and mode of evolution. *Paleobiology* 4, 120-134.

Borojevic, S. (1980) Can we develop varieties as we model them? *Proc. XIV. Int. Congr. Genet.* 1, 179-193.

Botstein, D. (1980) A theory of modular evolution for bacteriophages. *Ann. N.Y. Acad. Sci.* 354, 484-491.

Bowman, J.C. and Falconer, D.S. (1960) Inbreeding depression and heterosis of litter size in mice. *Genet. Res.* 1, 262-274.

Boyer, S.K. and Mallet, J.E. (1986) Characterization of *P. sativum* chloroplast psbA transcripts produced *in vitro, in vitro* and in *E. coli. Plant Mol. Biol.* 6, 229-243.

Brand, A., Breedan, L., Abraham, J., Sternglanz, R. and Nasmyth, K. (1985) Characterization of a 'silencer' in yeast: a DNA sequence with properties opposite to those of a transcriptional enhancer. *Cell* 41, 41-48.

Braun, W. (1965) *Bacterial Genetics.* 2nd ed. Saunders, Philadelphia.

Breitbart, R.E., Andreadis, A. and Nadal-Ginard, B. (1987) Alternative splicing: a ubiquitous mechanism for the generation of multiple protein isoforms from single genes. *Annu. Rev. Biochem.* <u>56</u>, 467-495.

Brian, M.V. (1965) Caste differentiation in social insects. *Symp. Zool. Soc. London* <u>41</u>, 13-38.

Briggs, R. (1972) Further studies on the maternal effect of the o gene in the Mexican axolotl, *J. Exp. Zool.* <u>181</u>, 271-280.

Britten, R.J. (1986) Rates of DNA sequence evolution differ between taxonomic groups. *Science Wash.* <u>231</u>, 1393-1398.

Britten, R.J. and Davidson, E.H. (1971) Repetitive and non-repetitive DNA sequences and a speculation on the origins of evolutionary novelty. *Q. Rev. Biol.* <u>46</u>, 111-113.

Britten R.J. and Davidson, E.H. (1976) DNA sequence arrangement and preliminary evidence on its evolution. *Fed. Proc.* <u>35</u>, 2151-2157.

Brittnacher, J.G., Sims, S.R. and Ayala, F.J. (1978) Genetic differentiation between species of the genus *Speyeria* (Lepidoptera; Nymphalidae). *Evolution* <u>32</u>, 199-210.

Brodkorb, P. (1971) Origin and evolution of birds. In *Avian Biology* Vol. 1, ed. D.S. Farner, J.R. King and K.C. Parkes, pp. 19-55. Academic Press, New York.

Brooks, J. and Shaw, G. (1973) *Origin and Development of Living Systems*. Academic Press, New York.

Brothers, A.J. (1976) Stable nuclear activation dependent on a protein synthesized during oogenesis. *Nature (London)* <u>260</u>, 112-115.

Brown, D.D. and Dawid, J.B. (1968) Specific gene amplification in oocytes. *Science Wash.* <u>160</u>, 272-280.

Brown, M.A., Glimchev, L.A., Nielsen, E.A., Paul, W.E. and Germain, R.W. (1986) T-cell recognition of Ia molecules selectively altered by a single amino acid substitution. *Science Wash.* <u>231</u>, 255-258.

Brown, K.I. and Nestor, K.E. (1974) Implications of selection for high and low adrenal response to stress *Poult Sci.* <u>53</u>, 1297-1302.

Bueg, H., Katz, F.E. and Gerisch, G. (1973) Dynamics of antigenic membrane sites relating to cell aggregation in *Dictyostelium discoideum*. *J. Cell. Biol.* <u>56</u>, 647-658.

Buetow, D.E. (1976) Phylogenetic origin of the chloroplast. *J. Protozool.* <u>23</u>, 41-47.

Bulling, E., Stephan, R. and Sebek, V. (1973) Die Entwicklung der Antibiotika-resistenz von Salmonellabakterien tierischer Herkunft in der Bundersrepublik Deutschland einschl. Berlin (West). 1. Mitteilung: Ein Vergleich zwischen 1961 and 1970/71. *Zentralbl. Bakteriol. Parasitenkd. Infektionskr. Hyg.* <u>225</u>, 245-256.

Cahn, P.H. (1959) Comparative optic development in *Astyanax mexicanus* and in two of its blind cave derivatives. *Bull. Am. Mus. Nat. Hist.* <u>115</u>, 72-112.

Cameron, J.R., Loh, E.Y. and Davis, R.W. (1979) Evidence for transcription of dispersed repetitive DNA families in yeast. *Cell* <u>16</u>, 739-751.

Campbell, K.S.W. and Marshall, C.R. (1987)   Rates of evolution among Palaeozoic echinoderms. In *Rates of Evolution*. ed. K.S.W. Campbell and M.F. Day. pp. 61-100. Allen and Unwin, London.

Campbell, R.D. and Porter, R.R.   (1983)   Molecular cloning and characterization of the gene coding for human complement protein factor. B. *Proc. Nat. Acad. Sci (U.S.A.)* 80, 4464-4468.

Carlile, M.J. (1975)   Taxes and tropisms: diversity, biological significance and evolution. In *Primitive Sensory and Communication Systems*. ed. M.J. Carlile pp. 1-28. Academic Press, London.

Carlson, M. and Botstein, D. (1982) Two differentially regulated mRNAs with different 5' ends encode secreted and intracellular forms of yeast invertase. *Cell* 28, 145-154.

Carpenter, F.M. (1977) Geological history and evolution of insects. *Proc XV. Congr. Entomol. Wash.* pp. 43-70.

Carrington, R. (1956)   *A Guide to Earth History*. Chatto and Windus, London.

Carson, H.L. (1975) The genetics of speciation at the diploid level. *Am. Nat.* 109, 83-92.

Cavalier-Smith, T. (1978) Nuclear volume control by nucleoskeletal DNA, selection for cell volume and cell growth rate, and the solution of the DNA C-value paradox. *J. Cell Sci.* 34, 247-278.

Cavalli-Sforza, L.L. and Maccacaro, G.A. (1952) Polygenic inheritance of drug-resistance in the bacterium *E. Coli. Heredity* 6, 311-331.

Chakravarti, A., Buetow, K.H., Antonarakis, S.E., Waber, P.G., Boehm, C.D. and Kazazian, H.H. (1984) Nonuniform recombination within the human ß-globin gene cluster. *Am. J. Hum. Genet.* 35, 1239-1258.

Chen, C.-H., Oishi, K.K., Kloeckener-Gruissem, A. and Freeling, M. (1987) Organ-specific expression of maize Adh-1 is altered after a *Mu* transposon insertion. *Genetics* 116, 469-477.

Chikaraishi, D.M., Deeb, S. and Sueoka, N. (1978) Sequence complexity of nuclear RNAs in adult rat tissues. *Cell* 13, 111-112.

Chopra, I. and Howe, T.G.B. (1978)   Bacterial resistance to the tetracyclines. *Microbiol. Rev.* 42, 707-724.

Clarke, B. (1970)   Selective constraints on amino-acid substitutions during the evolution of proteins. *Nature (London)* 228, 159-160.

Clarke, P.H. (1980) Experiments in microbial evolution: new enzymes, new metabolic activities. *Proc. R. Soc. London* B207, 385-404.

Clausen, J., Keck, D.A. and Hiesey, W.M. (1948) Experimental studies on the nature of species III. *Carnegie Inst. Washing Publ.* 581.

Clayton, G.A., Knight, G.R., Morris, J.A. and Robertson, A. (1957)   An experimental check on quantitative genetic theory III.   Correlated responses. *J. Genet.* 55, 171-180.

Cleland, R. (1974) Mode of action of plant hormones. In *Biochemistry of Hormones*. ed. H.V. Rickenberg, pp. 283-303. Butterworths, London.

Cloud, P. (1978)   *Cosmos, Earth and Man*. Yale University Press, New Haven.

Cloud, P. and Glaessner, M.F. (1982)   The Ediacarian period and system: the metazoa inherit the earth.   *Science Wash.* 218, 783-792.

Colbert, E.H. (1955)   *Evolution of the Vertebrates.* Wiley, New York.

Cooper, R., Herzog, C.E., Li, M.-L., Zapisek, W.F., Hoyt, P.R., Ratrie, H. and Papaconstantinou, J. (1984)   Localization of the LIMα family of repeated sequences in the mouse albumin-α-fetoprotein gene complex. *Nuc. Acids. Res.*12, 6575-6586.

Cope, E.D., (1896)   *The Primary Factors of Organic Evolution.* Open Court, Chicago.

Cox, C.G. and Gilbert, J.B. (1970)   Nonidentical times of gene expression in two strains of *Saccharomyces cerevisiae* with mapping differences. *Biochem. Biophys. Res. Commun.* 38, 750-756.

Craig, N.L. (1985)   Site-specific inversion: enhancers, recombination proteins and mechanism.   *Cell* 41, 649-650.

Craik, C.S., Rutter, W.J. and Fletterick, R. (1983)   Splice junctions: association with variation in protein structure, *Science Wash.* 220, 1125-1129.

Crawford, I.P. (1975)   Gene rearrangements in the evolution of the tryptophan synthetic pathway. *Bacteriol. Rev.* 39, 87-120.

Cronquist, A. (1968)   *The Evolution and Classification of Flowering Plants.* Nelson, London.

Crosa, J.H., Brenner, D.J., Ewing, W.H. and Falkow, S. (1973)   Nuclear relationships among the Salmonelleae.   *J. Bacteriol.*, 15, 307-315.

Crowson, R.A., Rolfe, W.D.I., Smart, J., Waterston, C.D., Willey, E.C. and Wootton, R.J. (1967)   Arthropoda: Chelicerata, Pycnogonida, Palaeoisopus, Myriopoda and Insecta.   In *The Fossil Record,* ed. W.B. Harland, C.H. Holland, M.R. House, N.F. Hughes, A.B. Reynolds, M.J.S. Rudwick, G.E. Slatterthwaite, L.B.H. Tarlo and E.C. Willey, pp. 499-534. Geological Society, London.

Cruz, V.F. de la., Lal, A.A., Welsch, J.A. and McCutchan, T.F. (1987) Evolution of the immuno-dominant domain of the circumsporozoite protein gene from *Plasmodium vivax.* *J. Biol. Chem.* 262, 6464-6467.

Cullimore, J.V., Gebhardt, C., Saarelainen, R., Miflin, B.J., Idler, K.B. and Barker, R.F. (1984)   Glutamine synthetase of *Phaseolus vulgaris* L. Organ-specific expression of a multigene family. *J. Mol. Appl. Genet.* 2, 589-599.

Cundliffe, E. (1980)   Antibiotics and prokaryotic ribosomes.   Action, interaction and resistance.   In *Ribosomes: Structure, Function and Genetics.* ed. G. Chambliss, G.R. Craven, J. Davies, K. Davis, I. Kahan and M. Nomura. pp. 555-581.   University Park Press, Baltimore.

Cunningham, E.P. (1984)   Present and future perspectives in animal breeding research. *Proc. XV. Int. Congr. Genet.* 4, 169-180.

Curtis, R., Macrina, F.L. and Falkinham, J.O. (1970)   *Escherichia coli* - an overview.   *Handb. Genet.* 1, 115-133.

Darwin, C. (1859)   *The Origin of Species.* Murray, London.

Davidse, L.C. (1982)   Benzimidazole compounds: selectivity and resistance.   In *Fungicide Resistance in Crop Protection,* ed. J. Dekker and G. Georgopolous, pp. 60-70. Centre for Agricultural Publishing, Wageningen.

Davidson, E.C. (1976) *Gene Activity in Early Development.* Academic Press, New York.

De Beer, G. (1958) *Embryos and Ancestors.* Clarendon Press, Oxford.

Decugis, H. (1941) *Le Vieillissement du Monde Vivant.* Masson, Paris.

Degani, C. and Halmann, M. (1967) Chemical evolution of carbohydrate metabolism. *Nature (London)* 216, 1207.

Dekker, J. (1986) Preventing and managing fungicide resistance. In *Pesticide Resistance. Strategies and Tactics for Management,* pp. 347-354, National Academy Press, Washington.

De Ley, J. and Kersters, K. (1975) Biochemical evolution in bacteria. In *Comprehensive Biochemistry* ed. M. Florkin and E.H. Stotz. Vol 29B, 1-77. Elsevier, Amsterdam.

De Lotto, R. and Spierer, P. (1986) A gene required for the specification of dorsal-ventral pattern in *Drosophila* appears to encode a serine protease. *Nature (London)* 323, 688-692.

Demerec, M. (1965) Homology and divergence in genetic material of *Salmonella typhimurium* and *Escherichia coli.* In *Evolving Genes and Proteins* ed. V. Bryson and H.J. Vogel, pp 505-510. Academic Press, New York.

Demerec, M. and New, K. (1965) Genetic divergence in *Salmonella typhimurium, S. montevideo* and *Escherichia coli. Biochem. Biophys. Res. Commun.* 18, 652-655.

Desplan, C., Theis, J. and O'Farrell, P.H. (1985) The *Drosophila* developmental gene, engrailed encodes a sequence-specific DNA binding activity. *Nature (London)* 318, 630-635.

Devilliers, C. (1965) The role of morphogenesis in the origin of higher levels of organization. *Syst. Zool.* 14, 259-271.

Devonshire, A.L. and Sawicki, R.M. (1979) Insecticide resistant *Myzus persicae* as an example of evolution by gene duplication. *Nature (London)* 280, 140-141.

Dhouailly, D. and Sengel, P. (1973) Interactions morphogènes entre l'epiderme de reptile et le derme d' oiseau ou de mammifère. *C.R. Acad. Sci D* 277, 1221-1224.

Dickerson, G.E. and Willham, R.L. (1983) Quantitative genetic engineering of more efficient animal production. *J. Mol. Evol.* 16, 73-94.

Dickerson, R.E. (1971) The structure of cytochrome C and the rates of molecular evolution. *J. Mol. Evol.* 1, 26-45.

Dickerson, W.J. (1980) Evolution of patterns of gene expression in Hawaiian picture-winged *Drosophila. J. Mol. Evol.* 16, 73-94.

Dilcher, D.L. (1974) Approaches to the identification of angiosperm leaf remains. *Bot. Rev.* 40, 1-157.

Doane, W.W. (1973) Role of hormones in insect development. In *Developmental Systems: Insects* ed. S.J. Counce and C.H. Waddington Vol. 2 pp. 291-498. Griffith, New York.

Domingo, E., Sabo, D., Taniguchi, T. and Weissman, C. (1978) Nucleotide sequence heterogeneity of an RNA phage population. *Cell* 13, 735-744.

Doolittle, R.F. (1985) The genealogy of some recently evolved vertebrate proteins. *Trends Biochem. Sci.* June pp. 233-237.

Doolittle, W.F. and Sapienza, C. (1980) Selfish genes, the phenotype paradigm and genome evolution. *Nature (London)* 284, 601-603.

Dover, G. (1982) Molecular drive: a cohesive mode of evolution. *Nature (London)* 299, 111-117.

Dower, N.A. and Stahl, F.W. (1981) χ activity during transduction-associated recombination. *Proc. Nat. Acad. Sci. USA* 78, 7033-7037.

Dowsett, A.P. and Young, M.W. (1982) Differing levels of dispersed repetitive DNA among closely related species of *Drosophila*. *Proc. Nat. Acad. Sci. USA* 79, 4570-4574.

Doyle, J.J., Schuler, M.A., Godette, W.D. Zenger, V., Beachy, R.N. and Slightom, J.L. (1986) The glycosylated seed protein storage proteins of *Glycine max* and *Phaseolus vulgaris*. *J. Biol. Chem.* 261, 9228-9238.

Drake, J.W. (1969) Comparative rates of spontaneous mutation. *Nature (London)* 221, 1132.

Dunlop, J.S.R., Muir, M.D., Milne, V.A. and Groves, D.I. (1978) A new micro-fossil assemblage from the Archaean of Western Australia. *Nature (London)* 254, 676-678.

Dunstone, R.L. and Evans, L.T. (1974) Role of changes in cell size in the evolution of wheat. *Aust. J. Plant. Physiol.* 1, 157-165.

Durham, J.W. (1971) The fossil record and the origin of the Deuterostomata. *Proc. N. Am. Paleontol. Conv.* part H, pp. 1104-1132.

Dusenberg, R.L. (1975) Characterization of the genome of *Phycomyces blakesleeanus*. *Biochim. Biophys. Acta* 378, 363-377.

Dutta, S.K. and Chaudhuri, R.K. (1975) Differential transcription of nonrepeated DNA during development of *Neurospora crassa*. *Dev. Biol.* 43, 35-41.

Dynan, W.S. (1986) Promoters for housekeeping genes. *Trends Genet.* August pp. 196-197.

Dynan, W.S. and Tjian R. (1985) Control of eukaryotic messenger RNA synthesis by sequence-specific DNA-binding proteins. *Nature (London)* 316, 774-778.

Dyte, C.E. (1967) Possible new approach to the chemical control of plant feeding insects. *Nature (London)* 216, 298.

Eakin, R.M. (1968) Evolution of photoreceptors. *Evol. Biol.* 12, 194-242.

Eck, R.V. and Dayhoff, M.O. (1966) Evolution of the structure of ferredoxin based on living relics of primitive amino acid sequences. *Science Wash.* 152, 363-366.

Edelman, G.M. (1983) Cell adhesion molecules. *Science Wash.* 219, 450-459.

Edlund, T., Grundström, T., and Normark, S. (1979) Isolation and characterization of DNA repetitions carrying the chromosomal ß-lactamase gene of *Escherichia coli* K-12. *Mol. Gen. Genet.* 173, 115-125.

Eibel, H. and Philippsen, P. (1984) Preferential integration of yeast transposable element Ty into a promoter region. *Nature (London)* 307, 386-388.

Eissenberg, J.C., Cartwright, J.L., Thomas, G.H. and Elgin, S.C.R. (1985) Selected topics in chromatin structure. *Annu. Rev. Genet.* 19, 485-536.

Elliott, H.C. (1970) *The Shape of Intelligence.* Allen and Unwin, London.

Emerson, B.M., Lewis, C.D. and Felsenfeld, G. (1985) Interaction of specific nuclear factors with the nuclear hypersensitive region of the chicken adult ß-globin gene: nature of the binding domain. *Cell* 41, 21-30.

Emmons, S.W. and Thomas, J.O. (1975) Tandem genetic duplications in phage lambda IV. The locations of spontaneously arising tandem duplication. *J. Mol. Biol.* 91, 147-152.

Epstein, C.J. (1970) The three-dimensional structure and evolution of proteins. In *Aspects of Protein Biosynthesis*, ed. C.B. Anfinsen, pp. 367-431. Academic Press, New York.

Erickson, J.M., Rahire, M., Bennoun, P., Delepelaire, P., Diner, B. and Rochaix, J. (1984) Herbicide resistance in *Chlamydomonas reinhardtii* results from a mutation in the chloroplast gene for the 32-kilodalton protein of photosystem II. *Proc. Nat. Acad. Sci. USA* 81, 3617-3621.

Evans, L.T. (1976) Physiological adaptation to performance in crop plants. *Philos. Trans. R. Soc. London* B275, 71-83.

Evans, L.T. (1980) The natural history of crop yield. *Am. Sci.* 68, 388-397.

Evans, L.T. (1983) Raising the yield potential: by selection or design? In *Genetic Engineering of Plants*, ed. T. Kosuge, C.P. Meredith and A. Hollaender, pp. 371-389. Plenum Press, New York.

Evans, L.T., Dunstone, R.L., Rawson, H.M. and Williams, R.F. (1970) The phloem of the wheat stem in relation to requirements for assimilate by the ear. *Aust. J. Biol. Sci.* 23, 743-752.

Eversole, D.E., Bergen, W.C., Merkel, R.A., Magee, W.T. and Harpster, H.W. (1981) Growth and muscle development of feedlot cattle of different genetic background. *J. Anim. Sci.* 53, 91-101.

Farabaugh, P.J. and Fink, G.R. (1980) Insertion of the eukaryotic transposable element TY 1 creates a 5-base pair duplication. *Nature (London)* 286, 352-356.

Favret, E.A., Favret, G.C. and Malvarex, E.M. (1975) Genetic regulatory mechanisms for seedling growth in barley. Barley genetics III. *Proc. 3rd Int. Congr. Barley Genet. Symp.* p. 37. Thiemig, Munich.

Feduccia, A. (1980) *The Age of Birds.* Harvard University Press, Massachusetts.

Feinstein, S.C., Ross, S.R. and Yamamoto, K.R. (1982) Chromosomal position effects determine transcriptional potential of integrated mammary tumor virus DNA. *J. Mol. Biol.* 156, 549-566.

Ferguson, E.L. and Horvitz, H.R. (1985) Identification and characterization of 22 genes that affect the vulval cell lineages of the nematode *Caenorhabditis elegans*. *Genetics* 110, 17-72.

Ferris, S.A. and Whitt, G.S. (1979)   Evolution of the differential regulation of duplicate genes after polyploidization *J. Mol. Evol.* 12, 267-317.

Field, L.J., Philbrick, W.M., Howles, P.N., Dickinson, D.P., McGowan, R.A. and Cross, K.W. (1984) Expression of tissue specific Ren-1 and Ren-2 genes of mice: comparative analysis of 5'-proximal flanking regions. *Mol. Cell Biol.* 4, 2321-2231.

Fink, G.R. (1987) Pseudogenes in yeast? *Cell* 49, 5-6.

Fink G.R., Boeke, J.D. and Garfinkel, D.J. (1986)   The mechanism and consequences of retrotransposition. *Trends Genet.* May 118-123.

Finnegan, D.J. (1981)   Transposable elements and proviruses. *Nature (London)* 292, 800-801.

Fisher, R.A. (1930) *The Genetic Theory of Natural Selection.* Clarendon Press, Oxford.

Fitch, W.M. and Atchley, W.R. (1985) Evolution in inbred strains of mice appears rapid. *Science Wash.* 228, 1169-1175.

Fitch, W.M. and Langley, C.H. (1976) Protein evolution and the molecular clock. *Fed. Proc.* 35, 2092-2097.

Fitz-James, P. and Young, E. (1969) Morphology of sporulation. In *The Bacterial Spore.* ed G. Gould and A. Hurst pp. 39-72. Academic Press, New York.

Fluhr, R., Kuhlemeier, C., Nagy, F. and Chua, N.H. (1986) Organ-specific and light-induced expression of plant genes. *Science Wash.* 232, 1106-1112.

Fogel, S. and Welch, J.W. (1984)   The molecular genetics of copper resistance in yeast: gene amplification via gene conversion. *Proc. XV. Int. Congr. Genet.* 2, 41-54.

Foster, T.J., Davis, M.A., Roberts, D.E., Takashita, K. and Klechner, M. (1981) Genetic organization of transposon Tn 10. *Cell* 23, 201-203.

Frankel, O.H., and Soulé, M.E. (1981)   *Conservation and Evolution.* Cambridge University press, Cambridge.

French, J. (1940)   Trial and error learning in *Paramecium.*   *J. Exp. Psychol.* 26, 609-615.

Freund, R. and Meselson, M. (1984) Long terminal repeat nucleotide sequence and specific insertion of the gypsy transposon. *Proc. Nat. Acad. Sci. USA* 81, 4462-4464.

Friedmann, J., Colwin, A.L. and Colwin, L.H. (1968)   Some structural aspects of fertilization in *Chlamydomonas reinhardi.*   *J. Cell Sci.* 3, 115-128.

Fuller, H.J. and Tippo, O. (1954) *College Botany.* Holt, New York.

Gale, M.D. and Law, C.N. (1977) The identification and exploitation of Norin 10 semi-dwarfing genes. *Annu. Rep. Plant Breeding Inst., Cambridge* p.21.

Gamble, C. (1980) Information exchange in the palaeolithic. *Nature (London)* 283, 522-523.

Garber, R.J. (1922) Inheritance and yield with particular reference to rust resistance and panicle type in oats. *Minn. Agr. Exp. Sta. Tech. Bull.* 7.

Gasser, C.S., Winter, J.A., Hironaka, C.M. and Shah, D.M. (1988) Structure, expression and evolution of the 5-enolpyruvylshikimate-3-phosphate synthase genes of *Petunia* and tomato. *J. Biol. Chem.* 263, 4280-4287.

Gellert, M., O'Dea, H.H., Itoh, T. and Tomizawa, J. (1976) Novobiocin and coumermycin inhibit DNA supercoiling catalyzed by DNA gyrase. *Proc. Nat. Acad. Sci. USA* 73, 4474-4478.

Georghiou, G.P. (1972) The evolution of resistance to pesticides. *Annu. Rev. Ecol. Syst.* 3, 133-168.

Georghiou, G.P. (1986) The magnitude of the resistance problem. In *Pesticide Resistance. Strategies and Tactics for Management.* pp. 14-43. National Academy Press, Washington.

Gilbert, L.I. and Frieden, E. (1981) *Metamorphosis.* Plenum Press, New York.

Gilbert, W. (1978) Why genes in pieces? *Nature (London) 271,* 501.

Gilbert, W. (1985) Genes-in-pieces revisited. *Science Wash.* 228, 823-824.

Gilbert, W. (1987) The exon theory of genes. *Cold Spr. Harb. Symp. Quant. Biol.* 52, 901-905.

Gilbert, W., Marchionni, M. and McKnight, G. (1986) On the antiquity of introns. *Cell* 46, 151-154.

Gillis, M., De Ley, J. and De Cleene, M. (1970) The determination of molecular weight of bacterial genome DNA from renaturation rates. *Eur. J. Biochem.* 12, 143-153.

Gingerich, P.D. (1976) Paleontology and phylogeny: patterns of evolution at the species level in early Tertiary mammals. *Am. J. Sci.* 276, 1-28.

Gladstones, J. (1967) Selection for economic characters in *Lupinus angustifolius* and *L. digitatus* 1. Non-shattering pods. *Aust. J. Exp. Agric. Anim. Husb.* 7, 360-366.

Goin, C.J., Goin, O.B. and Zug, G.R. (1978) *Introduction to Herpetology* 3rd ed. Freeman, San Francisco.

Goldberg, A.L. and Wittes, R.E. (1966) Genetic code: aspects of organization. *Science Wash.* 153, 420-424.

Goldschmidt, R. (1960) *The Material Basis of Evolution.* Pageant, New Jersey.

Good, R. (1956) *Features of Evolution in the Flowering Plants.* Longmans, London.

Goodman, M. Weiss, M.L. and Czelusniak, J. (1982) Molecular evolution above the species level: branching pattern, rates, and mechanisms. *Syst. Zool.* 31, 376-399.

Goudge, T.A. (1961) *The Ascent of Life.* Allen and Unwin, London.

Gould, S.J. (1977) *Ontogeny and Phylogeny.* Belknap Press, Massachusetts.

Grassé, P.-P. (1973) *L'Evolution du Vivant.* Editions Albin Michel, Paris.

Greenberg, M.J. (1959) Ancestors, embryos and symmetry. *Syst. Zool.* 8, 212-221.

Greenwood, P.H. (1973) Morphology, endemism and speciation in African cichlid fishes. *Verh. Deutsch Zool. Gesellsch.* <u>1973</u>, 115-124.

Gressel, J. and Segel, L.A. (1978) The paucity of plants evolving genetic resistance to herbicides: possible reasons and implications. *J. Theoret. Biol.* <u>75</u>, 349-371.

Greulach, V.A. (1973) *Plant Structure and Function.* Macmillan, New York.

Gribskov, M. and Burgess, R.R. (1986) Sigma factors from *E. coli, B. subtilis,* phage SP01 and phage T4 are homologous proteins. *Nuc. Acids. Res.* <u>14</u>, 6745.

Grosschedl, R. and Birnstiel, M.L. (1980) Spacer DNA sequences upstream of the TATAAATA sequence are essential for promotion of H2A histone gene transcription *in vivo. Proc. Nat. Acad. Sci. USA* <u>77</u>, 7102-7106.

Grüneberg, H. (1963) *The Pathology of Development, a Study of Inherited Skeletal Disorders in Animals.* Wiley, New York.

Gunderson, J.H., Elwood, H., Ingold, A., Kindle, K. and Sogin, M.L. (1987) Phylogenetic relationships between chlorophytes, chrysophytes and oomycetes. *Proc. Nat. Acad. Sci. USA* <u>84</u>, 5823-5827.

Gustafsson, A. (1947) Mutation in agricultural plants. *Hereditas* <u>33</u>, 1-100.

Haas, R. and Meyer, T.F. (1986) The repertoire of silent pilus genes in *Neisseria gonorrhoeae:* evidence for gene conversion. *Cell* 144, 107-115.

Hafen, E., Basler, K., Edstroem, J.-E. and Rubin, G.M. (1987) Sevenless, a cell-specific homeotic gene of *Drosophila,* encodes a putative transmembrane receptor with a tyrosine kinase domain. *Science (Wash.)* <u>236</u>, 55-63.

Hagedoorn, A.L. (1944) *Animal Breeding.* Lockwood, London.

Haldane, J.B.S. (1947) *Science Advances.* Allen and Unwin, London.

Haldane, J.B.S. (1954) The statics of evolution. In *Evolution as a Process,* ed. J. Huxley, A.C. Hardy and E.B. Ford, pp. 109-121. Allen and Unwin, London.

Hale, L.J. (1964) Cell movements, cell division and growth in the hydroid *Clytia johnstoni. J. Embryol. Exp. Morphol.* <u>12</u>, 517-538.

Halevy, A.H., Ashri, A. and Ben-Tal, Y (1969) Peanuts: gibberellin antagonists and genetically controlled differences in growth habit. *Science Wash.* <u>164</u>, 1397-1398.

Halvorson, H.O., Carter, B.L.A. and Tauro, P. (1971) Synthesis of enzymes during the cell cycle. *Adv. Microb. Physiol.* <u>6</u>, 47-106.

Hanley, B.A. and Schuler, M.A. (1988) Plant intron sequences: evidence for distinct groups of introns. *Nuc. Acids Res.* <u>16</u>, 7159-7176.

Hanson, E.D. (1976) Major evolutionary trends in animal protists. *J. Protozool.* <u>23</u>, 4-12.

Hanson, E.D. (1977) *The Origin and Early Evolution of Animals.* Pitman, London.

Harber, J.E. (1983) Mating type genes of *Saccharomyces cerevisiae*. In *Mobile Genetic Elements*, ed. J.A. Shapiro, pp. 559-561. Academic Press, New York.

Harlan, J.R. (1967) A wild wheat harvest in Turkey. *Archeology* 20, 197-201.

Harlan, J.R., de Wet, J.M.J. and Price, E.G. (1973) Comparative evolution of cereals. *Evolution* 27, 311-325.

Harland, W.B., Holland, C.H., House, M.R., Hughes, N.F., Reynolds, A.B., Rudwick, M.J.S., Slatterthwaite, G.E., Tarlo, L.B.H. and Willey, E.C. editors (1967) *The Fossil Record*. Geological Society, London.

Harlow, H.F. (1958) The evolution of learning. In *Behavior and Evolution*, ed. A. Roe and G.G. Simpson, pp. 269-290. Yale University Press, New Haven.

Hartl, D.L. and Dykhuisen, D.E. (1984) The population genetics of *Escherichia coli*. *Ann. Rev. Genet.* 18, 31-68.

Harvey, G. (1986) New hormone will transform Western dairying. *New Sci.* 2 Oct., p. 35.

Haseloff, J., Goelet, P., Zimmern, D., Ahlquist, P., Dasgupta, R. and Kaesberg, P. (1984) Striking similarities in amino acid sequence among nonstructural proteins encoded by RNA viruses that have dissimilar genetic organisation. *Proc. Nat. Acad. Sci. USA* 81, 4358-4362.

Hatzopoulos, A.K. and Regier, J.C. (1987) Evolutionary changes in the developmental expression of silkmoth chorion genes and their morphological consequences. *Proc. Nat. Acad. Sci. USA* 84, 479-483.

Heiser, C.B. (1973) *Seed to Civilization*. *The Story of Man's Food*. Freeman, San Francisco.

Hemmingsen, S.M., Woolford, C., van der View, S.M., Tilly, K., Dennis, D.T., Georgopoulos, C.P., Hendrix, R.W. and Ellis, R.J. (1988) Homologous plant and bacterial proteins chaperone oligomeric protein assembly. *Nature (London)* 333, 330-334.

Henning, M. (1975) Determination of cell shape in bacteria. *Annu. Rev. Microbiol.* 29, 45-60.

Herskowitz, J.H. (1967) *Basic Principles of Molecular Genetics*. Nelson, London.

Hickman, E.P., Hickman, C.P., Hickman, F.M. and Roberts, L.S. (1979) *Integrated Principles of Zoology*. Mosby, St. Louis.

Hilner, U. and Alberts, B.M. (1980) Fidelity of DNA replication catalyzed in vitro on a natural DNA template by the T4 bacteriophage multi-enzyme complex. *Nature (London)* 285, 300-305.

Hilu, K.W. (1983) The role of single-gene mutations in the evolution of flowering plants. *Evol. Biol.* 16, 97-128.

Hinde, R.A. (1966) *Animal Behaviour*. *A Synthesis of Ethology and Comparative Psychology*. McGraw-Hill, New York.

Hoffman, S., Chuong, C.-M. and Edelman, G.M. (1984) Evolutionary conservation of key structures and binding functions of neural cell adhesion molecules. *Proc. Nat. Acad. Sci. USA* 81, 6881-6885.

Holland, J., Spindler, K., Horodyski, F., Grabau, E., Nichol, S. and Vande Pol, S (1982) Rapid evolution of RNA genomes. *Science (Wash.)* <u>215</u>, 1577-1585.

Holland, P.W.H. and Hogan, B.L.M. (1986) Phylogenetic distribution of *Antennapedia*-like homoeo boxes. *Nature (London)* <u>321</u>, 251-253.

Holliday, R.J. and Putwain, P.D. (1980) Evolution of herbicide resistance in *Senecio vulgaris*: variation in susceptibility to simazine between and within populations. *J. Appl. Ecol.* <u>17</u>, 779-791.

Hollis, G.F., Hieter, P.A., McBride, O.W., Swan, D. and Leder, P. (1982) Processed genes: a dispersed human immunoglobulin gene bearing evidence of RNA-type processing. *Nature (London)* <u>296</u>, 321-325.

Holmes, N. and Parham, P. (1985) Exon shuffling *in vivo* can generate novel HLA class I molecules. *The EMBO J.* <u>4</u>, 2849-2854.

Holt, J.S., Stemler, A.J. and Redosevitch, S.R. (1981) Differential light responses of photosynthesis by triazine-resistant and triazine-susceptible *Senecio vulgaris* biotypes. *Plant Physiol.* <u>67</u>, 744-748.

Hood, L., Campbell, J.H. and Elgin, S.C.R. (1975) The organization, expression and evolution of antibody genes and other multigene families. *Annu. Rev. Genet.* <u>9</u>, 305-353.

Hopkinson, D.A., Edwards, Y.A. and Harris, H. (1976) The distribution of subunit numbers and subunit sizes of enzymes: a study of the products of 100 human gene loci. *Ann. Hum. Genet.* 383-411.

Hori, H. (1976) Molecular evolution of 5s RNA. *Mol. Gen. Genet.* <u>145</u>, 119-123.

Hori, H. and Osawa, S. (1987) Origin and evolution of organisms as deduced from 5s ribosomal RNA sequences. *Mol. Biol. Evol.* <u>4</u>, 445-472.

Hori, M., Kakihi, K. and Misato, T. (1977) Antagonistic effect of dipeptides on the uptake of polyoxin A by *Alternaria kikuchiana*. *J. Pestic. Sci.* <u>2</u>, 139-149.

Horowitz, S. and Gorovsky, M.A. (1985) An unusual genetic code in nuclear genes of *Tetrahymena*. *Proc. Nat. Acad. Sci.* <u>82</u>, 2452-2455.

Hoyer, B.H., McCarthy, B.J. and Bolton, E.T. (1964) A molecular approach to the systematics of higher organisms. *Science (Wash.)* <u>144</u>, 959-967.

Hrdlicka, A. (1929) Man's future in the light of the past and present. *Proc. Am. Phil. Soc.* <u>68</u>, 1-11.

Huxley, J. (1943) *Evolution. The Modern Synthesis.* Harper, New York.

Huxley, J. (1955) *Evolution in Action.* Chatto and Windus, London.

Irwin, D.M., Robertson, K.A. and MacGillivray, R.T.A. (1988) Structure and evolution of the bovine prothrombin gene. *J. Mol. Biol.* <u>200</u>, 31-45.

Ising, G. and Block, K. (1981) Derivation-dependent distribution of insertion rates for a *Drosophila* transposon. *Cold Spring Harbor Symp. Quant. Biol.* <u>45</u>, 527-551.

Jacobson, J.V. and Beach, L.R. (1985) Control of transcription of α-amylase and rRNA genes in barley aleurone protoplasts by gibberellin and abscisic acid. *Nature (London)* <u>316</u>, 275-277.

Jacquier, A. and Dujon, B. (1985)   An intron-encoded protein is active in a gene conversion process that spreads an intron into a mitochondrial gene. *Cell* 41, 383-394.

Jauniaux, J.C., Urrestarazu, L.A. and Wiame, J. (1978)   Arginine metabolism in *Saccharomyces cerevisiae*: subcellular localization of the enzymes. *J. Bacteriol.* 133, 1096/1107.

Jennings, P.A., Finch, J.T., Winter, G., and Robertson, J.S. (1983)   Does the higher order structure of the influenza virus ribonucleoprotein guide sequence rearrangements in influenza viral RNA? *Cell* 34, 619-627.

Jensen, R.A. (1976)   Enzyme recruitment in evolution of new function. *Annu. Rev. Microbiol.* 30, 409-425.

Jensen, R.A., Nasser, D.S. and Nester, E.W. (1967)   Comparative control of a branch-point enzyme in microorganisms. *J. Bacteriol.* 94, 1582-1583.

Jerison, H.J. (1970)   Brain evolution: New light on old principles. *Science (Wash)* 170, 1224-1225.

Jerison, H.J. (1973)   *Evolution of the Brain and Intelligence.* Academic Press, New York.

Jinks, J.L. (1981)   The genetic framework of plant breeding.   *Phil. Trans. R. Soc. Lond.* B 292, 407-419.

Jung, G., Korn, E.D. and Hammer, J.A. (1987)   The heavy chain of *Acanthamoeba* myosin 1B is a fusion of myosin-like and non-myosin-like sequences. *Proc. Nat. Acad. Sci. USA* 84, 6720-6724.

Kafatos, F.C. and 16 others (1977)   *Results and Problems in Cell Differentiation* 8, 45-145. Springer, Berlin.

Kaiser, D. (1986)   Control of multicellular development. *Dictyostelium* and *Myxococcus. Annu. Rev. Genet.* 20, 539-566.

Kamp, D., Kahmann, R., Zipser, D., Broker, D.R. and Chow, L.T. (1978)   Inversion of the G DNA segment of phage Mu controls phage specificity. *Nature (London)* 271, 577-580.

Kandel, E.R. and Schwartz, J.H. (1982)   Molecular biology of learning: modulation of transmitter release. *Science (Wash)* 218, 433-443.

Kaplan, L. (1965)   Archeology and domestication in American *Phaseolus* (beans). *Econ. Bot.* 19, 358-368.

Katinakis, P. and Verma, D.P.S. (1985)   Nodulin-24 gene of soybean codes for a peptide of the peribacteroid membrane and was generated by tandem duplication of a sequence resembling an insertion element. *Proc. Nat. Acad. Sci USA* 82, 4157-4161.

Kearsey, S.E. and Craig, I.W. (1981)   Altered ribosomal RNA genes in mitochondria from mammalian cells with chloramphenicol resistance. *Nature (London)* 290, 607-608/

Keese, P. and Symons, R.H. (1985)   Domains in viroids: Evidence of intermolecular RNA rearrangements and their contribution to viral evolution. *Proc. Nat. Acad. Sci. USA* 82, 4582-4586.

Kettler, M.K., Ghent, A.W. and Whitt, G.S. (1986)   A comparison of phylogenies based on structural and tissue-expressional differences of enzymes in a family of teleost fishes (Salmoniformes, Umbridae). *Mol. Biol. Evol.* <u>3</u>, 485–498.

Kidd, S., Kelley, M.R. and Young, M.W. (1986) Sequence of the Notch locus of *Drosophila melanogaster:* relationship of the encoded protein to mammalian clotting and growth factors. *Mol. Cell Biol.* <u>6</u>, 3094–3108.

Kiehn, E.D. and Holland, J.J. (1970) Size distribution of polypeptide chains in cells. *Nature (London)* <u>226</u>, 544–545.

Kimmel, A.R. and Firtel, R.A. (1983) Sequence organization in *Dictyostelium:* unique structure at the 5'-end of protein coding genes. *Nuc. Acids Res.* <u>11</u>, 541–552.

Kimura, M. (1985) The neutral theory of molecular evolution. *New Scientist* 11 July, pp. 41–44.

King, M. and Wilson, A.C. (1975) Evolution at two levels in humans and chimpanzees. *Science (Wash)* <u>188</u>, 107–116.

King, R.J. and Mainwaring, W. (1974) *Steroid-Cell Interactions.* Butterworths, London.

Kirkpatrick, M. and Selander, R.K. (1979) Genetics of speciation in lake white fishes in the Allegash basin. *Evolution* <u>33</u>, 478–485.

Kiser, J.S., Gale, G.O. and Kemp, G.A. (1969) Resistance to antimicrobial agents. *Adv. Appl. Microbiol.* <u>11</u>, 77–100.

Kleene, K.C. and Humphreys, T. (1977) Similarity of hn RNA sequences in blastula and pluteus stage sea urchin embryos. *Cell* <u>12</u>, 143–155.

Klein, H.L. and Petes, T.D. (1981) Intrachromosomal gene conversion in yeast. *Nature (London)* <u>239</u>, 144–148.

Kochert, G. (1973) Colony differentiation in green algae. In *Developmental Regulation* ed. S.J. Coward pp. 155–167. Academic Press, New York.

Kopecko, D.J. (1980) Specialized genetic recombination systems in bacteria and their involvement in gene expression and evolution. *Prog. Mol. Subcell. Biol.* <u>7</u>, 135–234.

Krieger, R.I., Feeny, P.P. and Wilkinson, C.F. (1971) Detoxication enzymes in the guts of caterpillars: an evolutionary answer to plant defences? *Science (Wash)* <u>172</u>, 579–581.

Kronstad, J.W. and Whiteley, H.R. (1984) Inverted repeat sequences flank a *Bacillus thuringiensis* crystal protein gene. *J. Bacteriol.* <u>160</u>, 95–102.

Kuff, E.L., Feenstra, A., Lueders, K., Rechavi, G., Givol, D. and Canaani, E. (1983) Homology between an endogenous viral LTR and sequences inserted in an activated cellular oncogene. *Nature (London)* <u>302</u>, 547–548.

Kunkel, T.A. and Loeb, L.A. (1980) On the fidelity of DNA replication. The accuracy of *Escherichia coli* DNA polymerase I in copying natural DNA *in vitro*. *J. Biol. Chem.* <u>225</u>, 9961–9966.

Kunkel, T.A. and Loeb, L.A. (1981) Fidelity of mammalian DNA polymerases. *Science (Wash)* <u>213</u>, 765–767.

Kurtén, G. (1959) Rates of evolution in fossil mammals. *Cold Spring Harbor Symp. Quant. Biol.* 24, 205-215.

Kurtén, B. (1969) Continental drift and evolution. *Sci. Am.* 220, 54-64.

Laimins, L., Holmgren-König, M. and Khoury, G. (1986) Transcriptional 'silencer' element in rat repetition sequences associated with the rat insulin 1 gene locus. *Proc. Nat. Acad. Sci. USA* 83, 3151-3155.

Land, R.B. and Carr, W.R. (1979) Reproduction in domestic animals. In *Genetic Variation in Hormone Systems.* ed. T.G.M. Shire. C.R.C. Press, West Palm Beach.

Langridge, J. (1969) Mutations conferring quantitative and qualitative increases in ß-galactosidase activity in *Escherichia coli*. *Mol. Gen. Genet.* 105, 74-83.

Langridge, J. (1974) Mutation spectra and the neutrality of mutations. *Aust. J. Biol. Sci.* 27, 309-319.

Langridge, J. (1982) Precambrian evolutionary genetics. In *Mineral Deposits and the Evolution of the Biosphere.* ed. H.D. Holland and M. Schidlowski. pp. 83-102. Dahlem Konferenzen, Springer, Berlin.

Langridge, J. (1987) Old and new theories of evolution. In *Rates of Evolution.* ed. K.W.S. Campbell and M.F. Day, pp. 248-262. Allen and Unwin, London.

Larsell, O. (1963) Comparative neurology. *Encyl. Brit.* 16, 256-257.

Larson, A. and Highton, R. (1978) Geographical protein variation and divergence in the salamanders of the *Plethodon welleri* group (Amphibia, Plethodontidae). *Syst. Zool.* 27, 431-448.

Laurent, M., Pays, E., Delinte, K., Magnus, E., Van Meirvenne, N. and Steinert, M. (1984) Evolution of a trypanosome surface antigen gene repertoire linked to non-duplicative gene activation. *Nature (London)* 308, 370-373.

Lawlor, T.E. (1979) *Handbook to the Orders and Families of Living Mammals.* Mad River Press, California.

Ledley, F.D., Dililla, A.G., Kwok, C.M. and Woo, S.L.C. (1985) Homology between phenylalanine and tyrosine hydrolases reveals common structural and functional domains. *Biochemistry* 24, 3389-3394.

Lees, A.D. (1966) The control of polymorphism in aphids. *Adv. Insect Physiol.* 3, 277-279.

Levin, D.A. and Wilson, A.C. (1976) Rates of evolution in seed plants: Net increase in diversity of chromosome numbers and species through time. *Proc. Nat. Acad. Sci. USA* 73, 2086-2090.

Levin, R. (1983) How mammalian RNA returns to the genome. *Science (Wash.)* 219, 1052-1054.

Levis, R., Dunsmuir,P. and Rubin, G.M. (1980) Terminal repeats of the *Drosophila* transposable element copia: nucleotide and genomic organization. *Cell* 21, 581-588.

Lewis, W.H. (1980) *Polyploidy: Biological Relevance.* Plenum Press, New York.

Li, W., Gojobori, T. and Nei, N. (1981) Pseudogenes as a paradigm of neutral evolution. *Nature (London)* 1326, 93-96.

Li, W.-H. and Tanimura, M. (1987) The molecular clock runs more slowly in man than in apes and monkeys. *Nature (London)* 326, 93-96.

Liem, K.F. (1974) Evolutionary strategies and morphological innovations: cichlid pharyngeal jaws. *Syst. Zool.* 22, 425-441.

Lieth, H. and Whittaker, R.H. (Editors) 1975) *Ecological studies: analysis and synthesis* Vol. 14. Primary productivity of the biosphere, p. 306. Springer, Berlin.

Loeb, L.A. and Kunkel, T.A. (1982) Fidelity of DNA synthesis. *Annu. Rev. Biochem.* 52, 429-457.

Loeblich, A.R. (1974) Protisten phylogeny as indicated by the fossil record. *Taxon* 23, 277-290.

Loomis, W.F. (1982) *The Development of Dictyostelium discoideum.* Academic Press, New York.

Lorenz, K. (1941) Vergleichende Bewegungsstudien an Anatinen. *J. Ornithol.* 89, (suppl.) 194-294.

Losick, R. and Pero, J. (1981) Cascades of sigma factors. *Cell* 25, 582-584.

Lueders, K., Leder, A., Leder, P. and Kuff, E. (1982) Association between a transposed α-globin pseudogene and retrovirus-like elements in the BALB/c mouse genome. *Nature (London)* 295, 426-428.

MacLean, N. (1976) *Control of Gene Expression.* Academic Press, London.

McCahon, D. and Slade, W.R. (1981) A sensitive method for the detection and isolation of recombinants of foot- and mouth-disease virus. *J. Gen. Virol.* 53, 333-342.

McGhee, J.D. and Felsenfeld, G. (1980) Nucleosome structure. *Annu. Rev. Biochem.* 49, 1115-1156.

McGinnis, W., Garber, R.L., Wirz, J., Kuroiwa, A. and Gehring, W.J. (1984) A homologous protein coding sequence in *Drosophila* homeotic genes and its conservation in other metazoans. *Cell* 37, 403-408.

McKenzie, J.A., Dearn, J.M. and Whitten, M.J. (1980) Genetic basis of resistance to diazinon in Victorian populations of the Australian sheep blowfly, *Lucilia cuprina*. *Aust. J. Biol. Sci.* 33, 85-95.

McKnight, G.L., O'Hara, P.J. and Parker, M.L. (1986) Nucleotide sequences of the triose phosphate isomerase gene from *Aspergillus nidulans*: implications for a differential loss of introns. *Cell* 46, 114-147.

Maeda, N., Bliska, J.B. and Smithies, O. (1983) Recombination and balanced chromosome polymorphism suggested by DNA sequences 5′ to the human δ globin gene. *Proc. Nat. Acad. Sci. USA* 80, 5012-5016.

Maeda, N. and Smithies, O. (1986) The evolution of multigene families: Human haptoglobin genes. *Annu. Rev. Genet.* 20, 81-108.

Maeda, N., Yang, F., Barnett, D. R., Bowman, B.H. and Smithies, O. (1984) Duplication within the haptoglobin Hp^2 gene. *Nature (London)* 309, 131-135.

Magasanik, B. (1982) Genetic control of nitrogen assimilation in bacteria. *Annu. Rev. Genet.* 16, 135-168.

Mangelsdorf, P.C. (1975) Quoted in Harlan, J.R. (1975) - *Crops and Man.* American Society of Agronomy, Wisconsin.

198

Margulis, L. (1970) *Origin of Eukaryotic Cells.* Yale University Press, New Haven.

Margulis, L. and Sagan, D. (1986) *Origins of Sex.* Yale University Press, New Haven.

Markert, C.L. and Ursprung, H. (1971) *Developmental Genetics.* Prentice-Hall, New Jersey.

Martinelli, S.D. and Clutterbuck. A.J. (1971) A quantitative survey of conidiation mutants in *Aspergillis nidulans.* *J. Gen. Microbiol.* 69, 262-268.

May, R.M. and Dobson, A.P. (1986) Population dynamics and the rate of evolution of pesticide resistance. In *Pesticide Resistance. Strategies and Tactics for Management.* pp. 170-193. National Academy Press, Washington.

Mayr, E. (1960) The emergence of evolutionary novelty. In *Evolution after Darwin.* ed. S. Tax.1 Vol. 1, pp. 349-380. University of Chicago Press, Chicago.

Mayr, E. (1963) *Animal Species and Evolution.* Harvard University Press, Massachusetts.

Mayr, E. (1967) *Populations, Species and Evolution.* Belknap Press, Massachusetts.

Medawar, P.R. (1967) *The Art of the Soluble.* Methuen, London.

Meier, J.T., Simon, M.I. and Barbour, A.G. (1985) Antigenic variation is associated with DNA rearrangements in a relapsing fever *Borrelia. Cell* 41, 403-409.

Mertins, P. and Gallwitz, D. (1987) Nuclear pre-mRNA splicing in the fission yeast *Schizosaccharomyces pombe* strictly requires an intron-contained conserved sequence element. *EMBO J.* 6, 1757-1763.

Meyer, T.F., Billyard, E., Haas, R., Storzbuch, S. and So, M. (1984) Pilus genes of *Neisseria gonorrhoeae*: chromosomal organization and DNA sequence. *Proc. Nat. Acad. Sci. USA* 81, 6110-6114.

Michel, F. and Lang, B.F. (1985) Mitochondrial class II introns encode proteins related to the reverse transcriptases of retroviruses. *Nature (London)* 316, 641-643.

Michels, P.A.M. (1986) Evolutionary aspects of trypanosomes: analysis of genes. *J. Mol. Evol.* 24, 45-52.

Michelson, A.M., Blake, C.F., Evans, S.T. and Orkin, S.H. (1985) Structure of the human phosphoglycerate kinase gene and the intron-mediated evolution and dispersal of the nucleotide-binding domain. *Proc. Nat. Acad. Sci. USA* 82, 6965-6969.

Miklos, G.L.G. and Gill, A.C. (1981) The DNA sequences of cloned complex satellite DNAs for Hawaiian *Drosophila* and their bearing on satellite DNA sequence conservation. *Chromosoma* 82, 409-427.

Milkman, R.D. (1961) The genetic basis of natural variation III. Developmental lability and evolutionary potential. *Genetics* 46, 25-38.

Miller, C.G. and Roth, J.R. (1971) Recessive-lethal nonsense suppressors in *Salmonella typhimurium.* *J. Mol. Biol.* 59, 63-75.

199

Miller, J., McLachlan, A.D. and King, A. (1985)   Repetitive zinc-binding domains in the protein transcription factor IIIA from *Xenopus* oocytes. *EMBO J.* <u>4</u>, 1609-1614.

Miller, S.L. and Orgel, L.E. (1974)   *The Origins of Life on Earth*. Prentice-Hall, New Jersey.

Minchin, E.A. (1912)   *An Introduction to the Study of the Protozoa*. Arnold, London.

Moorbath, S., O'Nions, R.K. and Pankhurst, R.J. (1973)   Early Archaean age for the Isua iron formation, West Greenland. *Nature (London)* <u>245</u>, 138-139.

Moore, G.P. (1983)   Slipped-mispairing and the evolution of introns. *Trends Biochem. Sci.* November p. 411-414.

Morowitz, H.J. (1966)   The minimum size of cells.   In *Principles of Biomolecular Organization*. ed. G.E.W. Wolstenholm and M. O'Connor. pp. 446-462. Ciba Foundation Symposium, London.

Morowitz, H. (1971)   Energy flow and biological organization (paper presented at the *International Biophysics Meeting*, Sept. 1968, Cambridge, Massachusetts).

Mortlock, R.P. (1983)   Experiments in evolution using microorganisms. *BioScience* <u>33</u>, 308-313.

Moseley, W.M., Krabill, L.F. and Olsen, R.F. (1982)   Effect of bovine growth hormone administered in various patterns in nitrogen metabolism in the Holstein steer.   *J. Anim. Sci.1* <u>55</u>, 1062-1070.

Muggeo, M., Ginsberg, B.H., Roth, J., Neville, D.M., De Mevis, P. and Kahon, R. (1979) The insulin receptor in vertebrates is functionally more conserved during evolution than insulin itself.   *Endocrinology* <u>104</u>, 1393-1397.

Müller, A.H.   (1955)   *Der Grossablauf der stammesgeschichtlichen Entwicklung*. Fischer, Jena.

Muller, J. (1981) Fossil pollen records of extant angiosperms. *Bot. Rev.* <u>47</u>, 1-142.

Müller, W.E.G., Müller, I., Zahn, and Kurelee, B. (1976)   Species-specific aggregation factor in sponges VI. Aggregation receptor from the cell surface.   *J. Cell Sci. 21*, 227-241.

Mulvihill, E.R., Le Pennec, J.P. and Chambon, P. (1982) Chicken oviduct progesterone receptor: location of specific region of high affinity binding in cloned DNA.   *Cell* <u>24</u>, 621-632.

Myers, R.M., Tilly, K. and Maniatus, T. (1986)   Fine structure genetic analysis of a ß-globin promoter.   *Science (Wash.)* <u>232</u>, 613-618.

Nagl, W. (1979) Nuclear ultrastructure: condensed chromatin in plants is species-specific (karyotypical), but not tissue-specific (functional). *Protoplasma* <u>100</u>, 53-71.

Nagy, F., Morelli, G., Fraley, R.T., Rogers, S.G. and Chua, N.-H. (1985) Photo-regulated expression of a pea rbc S gene in leaves of transgenic plants. *EMBO J.* <u>12</u>, 3063-3968.

Nahmias, A.J. and Reanney, D.C. (1977) The evolution of viruses. *Annu. Rev. Ecol. Syst.* <u>8</u>, 29-49.

Nanney, D.L. (1982)   Genes and phenes in *Tetrahymena*.   *BioScience* 32, 783-788.

Neidhardt, F.C., Van Bogelen, R.A. and Vaughan, V. (1984)   The genetics and regulation of heat-shock proteins. *Annu. Rev. Genet.* 18, 295-329.

Nelson, D.R. and Zusman, D.R. (1983)   Evidence for long-lived mRNA during fruiting body formation in *Myxococcus xanthus*.   *Proc. Nat. Acad. Sci. USA* 80, 1467-1471.

Nicol, D., Desborough, G.A. and Solliday, J.R. (1959)   Paleontological record of the primary differentiation in some major invertebrate groups. *J. Wash. Acad. Sci.* 49, 351-366.

Nissen, H.W. (1955)   Problems of mental evolution in the primates.   In *The Non-human Primates and Human Evolution.* ed. J.A. Gavan, pp. 99-109, Wayne University Press, Detroit.

Nolan, J. (1986)   Mechanisms of insecticide resistance.   In *Insect Resistance Management.* Ed. J.C. Daly and D.M. Suckling, pp. 34-39. CSIRO Site Services, Canberra.

Nomura, M. and Morgan, E.A. (1977)   Genetics of bacterial ribosomes. *Annu. Rev. Genet.* 11, 297-347.

Nordström, K., Ingram, L.C. and Lundbäck, A. (1972)   Mutations in R factors of *Escherichia coli* causing an increased number of R factor copies per chromosome. *J. Bacteriol.* 110, 562-569.

Nüsslein-Volhard, C. and Wieschaus, E. (1980)   Mutations affecting segment number and polarity in *Drosophila*. *Nature (London)* 287, 795-801.

O'Brien, S.J., Roelke, M.E., Marker, L., Winkler, C.A., Meltzer, D., Colly, L., Evermann, J.F., Bush, M. and Wildt, D.E. (1985)   Genetic basis for species vulnerability in the cheetah. *Science Wash.* 227, 1428-1434.

Ochman, H. and Selander, R.K. (1984)   Evidence for clonal population structure in *Escherichia coli*. *Proc. Nat. Acad. Sci. USA* 81, 198-201.

Ochman, H. and Wilson, A.C. (1987)   Evolution in bacteria: evidence for a universal substitution rate in cellular genomes. *J. Mol. Evol.* 26, 74-86.

O'Connor, M.B., Binari, R., Perkins, L.A. and Bender, W. (1988)   Alternative RNA products from the Ultrabithorax domain of the bithorax complex. *EMBO J.* 7, 435-445.

Ohno, S. (1969) The preferential activation of maternally derived alleles in development of interspecific hybrids.   In *Heterospecific Genome Interactions* ed. V. Defendi pp. 137-150.   Wistar Institute Press, Philadelphia.

Ohno, S. (1970) *Evolution by Gene Duplication.* Allen and Unwin, London.

Ohyama, K., Kohchi, T., Sano, T. and Yamada, Y. (1988)   Newly identified groups of genes in chloroplasts. *Trends Biochem. Sci.* Jan 1988, pp. 19-22.

Olson, E. (1970)   *Vertebrate Paleozoology.* Wiley-Interscience, New York.

Olson, J.M. (1978)   Precambrian evolution of photosynthetic and respiratory organisms.   In *Evolutionary Biology.* ed. T. Dobzhansky, M.K. Hecht and W.C. Steere 11, 1-37.

Olson, J.M. and Pierson, B.K. (1986) Photosynthesis 3.5 thousand million years ago. *Photosynthesis Res.* <u>9</u>, 251-259.

Olson, M.O.J., Nagabhushan, N., Dzwiniel, M. and Smillie, L.B. (1970) Primary structure of α-lytic protease: a bacterial homologue of the pancreatic proteases. *Nature (London)* <u>228</u>, 438-442.

Orgel, L.E. and Crick, F.H.C. (1980) Selfish DNA: the ultimate parasite. *Nature (London)* <u>284</u>, 604-607.

Ornston, L.N. and Parke, D. (1977) The evolution of induction mechanisms in bacteria: insights derived from the study of the ß-ketoadipate pathway. *Curr. Topics Cell Reguln.* <u>12</u>, 209-262.

Ornston, L.N. and Stanier, R.Y. (1966) The conversion of catechol and protocatechuate to ß-ketoadipate by *Pseudomonas putida*. *J. Biol. Chem.* <u>241</u>, 3776-3786.

Osiewacz, H.D. and Esser, K. (1984) The mitochondrial plasmid of *Podospera anserina*: a mobile intron of a mitochondrial gene. *Curr. Genet.* <u>8</u>, 299-305.

Osinga, K.A., Swinkels, B.W., Gibson, W.C., Borst, P., Vaeneman, G.H., Van Bloom, J.H., Michels, P.A.M. and Opperdoes, F.R. (1985) Topogenesis of microbody enzymes: a sequence comparison of the genes for the glycosomal (microbody) and cytosolic phosphoglycerate kinases of *Trypanosoma brucei*. *EMBO J.* <u>4</u>, 3811-3817.

Ottea, J.A. and Plapp, F.W. (1984) Glutathione-S-transferase in the house fly. Biochemical and genetic changes associated with induction and insecticide resistance. *Pestic. Biochem. Physiol.* <u>22</u>, 203-208.

Oullette, M., Bissonette, L. and Roy, P.H. (1987) Precise insertion of antibiotic resistance determinants into Tn 21-like transposons: Nucleotide sequence of the OXA-1 ß-lactamase gene. *Proc. Nat. Acad. Sci USA* <u>84</u>, 7378-7382.

Paigen, K. (1979) Acid hydrolases as models of genetic control. *Annu. Rev. Genet.* <u>13</u>, 417-466.

Paigen, K. (1986) Gene regulation and its role in evolutionary processes. In *Evolutionary Processes and Theory*, ed. S. Karlin and E. Nevo, pp. 3-36. Academic Press, New York.

Palmer, A.R. (1983) The decade of North American geology. 1983. Geological time scale. *Geology* <u>11</u>, 503-504.

Palmiter, R.D. and Brinster, R.L. (1985) Transgenic mice. *Cell* <u>41</u>, 343-345.

Parker, H.R., Philipp, D.P. and Whitt, G.S. (1985) Gene regulatory divergence among species estimated by altered developmental patterns in interspecific hybrids. *Mol. Biol. Evol.* <u>2</u>, 217-250.

Pasteels, J. (1940) Un aperçu comparatif de la gastrulation chez les chordes. *Biol. Rev.* <u>15</u>, 59-106.

Patthy, L. (1985) Evolution of the proteases of blood coagulation and fibrinolysis by assembly from modules. *Cell* <u>41</u>, 657-663.

Patthy, L. (1987) Intron-dependent evolution: preferred types of exons and introns. *FEBS Lett.* <u>214</u>: 1-7.

Paulus, H. (1983) The evolutionary history of the ornithine cycle as a determinant of its structure and regulation. *Curr. Top. Cell Regul.* 22, 177-200.

Paulus, T.J. and Switzer, R.L. (1979) Characterization of pyrimidine-repressible and arginine-repressible carbamyl phosphate synthetases from *Bacillus subtilis*. *J. Bacteriol.* 137, 82-91.

Peacock, F.C. (1976) Growth regulation issue. In *Outlook on Agriculture.* Vol. 9, I.C.I., Bracknell, England.

Pearston, D.H., Gordon, M. and Hardman, N. (1985) Transposon-like properties of the major, long repetitive sequence family in the genome of *Physarum polycephalum*. *EMBO J.* 4, 3557-3562.

Pelham, H.R.B. (1982) A regulatory up-stream promoter in the *Drosophila* hsp 70 heat shock gene. *Cell* 30, 517-528.

Perutz, M.F. (1983) Species adaptation in a protein molecule. *Mol. Biol. Evol.* 1, 1-28.

Perutz, M.F. and Lehmann, H. (1968) Molecular pathology of human haemoglobin. *Nature (London)* 219, 902-909.

Pickersgill, B. and Heiser, C.B. (1976) Cytogenetics and evolutionary change under domestication. *Phil. Trans. Roy. Soc.* B275, 55-69.

Pickett-Heaps, J.D. (1974) Cell division and evolution and branching in *Oedocladium* (Chlorophyceae). *Cytobiologie* 14, 319-337.

Piggot, P.J. (1985) Sporulation of *Bacillus subtilis*. In *Molecular Biology of the Bacilli*, ed. D. Dubnau 2 pp. 74-108. Academic Press, New York.

Plapp, F.W. (1984) The genetic basis of insecticide resistance in the housefly. Evidence that a single locus plays a major role in metabolic resistance to insecticides. *Pestic. Biochem. Physiol.* 22, 194-201.

Plasterk, R.H.A., Brinkman, A. and van de Platte, P. (1983) DNA inversions in the chromosome of *Escherichia coli* and in bacteriophage Mu: relationship to other site-specific recombination systems. *Proc. Nat. Acad. Sci. USA* 80, 5355-5358.

Poethig, R.S., Coe, E.H. and Johri, M.M. (1986) Cell lineage patterns in maize embryogenesis: a clonal analysis. *Dev. Biol.* 117, 392-404.

Poodry, C.A., Hall, L. and Suzuki, D.T. (1973) Developmental properties of 'chabira [ts]', a pleiotropic mutation affecting larval and adult locomotion and development. *Dev. Biol.* 32, 378-388.

Prat, A., Katinka, M., Caron, F. and Meyer, E. (1986) Nucleotide sequence of the *Paramecium primaurelia* G surface protein. *J. Mol. Biol.* 189, 47-60.

Quigley, F., Martin, W.F. and Cerff, R. (1988) Intron conservation across the prokaryote-eukaryote boundary: structure of the nuclear gene for chloroplast glyceraldehyde-3-phosphate dehydrogenase from maize. *Proc. Nat. Acad. Sci. USA.* 85, 2672-2676.

Raff, R.A. and Kaufman, T.C. (1983) *Embryos, Genes and Evolution.* Macmillan, New York.

203

Raff, R.A. and Raff, E.C. (1970) Respiratory mechanisms and the metazoan fossil record. *Nature (London)* 228, 1003-1005.

Raven, H.C. (1940) On the anatomy and evolution of the locomotor apparatus of the nipple-tailed ocean sunfish *(Masturus lanceolatus).* *Bull. Am. Mus. Nat. Hist.* 76, 143-150.

Reanney, D. (1984) Genetic noise in evolution? *Nature (London)* 307, 318-319.

Rees, H. (1972) DNA in higher plants. *Brookhaven Symp. Biol.* 23, 394-418.

Rendel, J.M. (1984) Decline in the number of breeds, its consequences and remedies. *Proc. XV. Int. Congr. Genet.* 4, 23-33.

Rensch, B. (1959) *Evolution above the Species Level.* Methuen, London.

Riek, E.F. (1970) Fossil history. In *The Insects of Australia.* pp. 168-186. CSIRO, Melbourne University Press, Melbourne.

Riley, R. (1980) Genetics and world grain production. *Proc. XIV. Int. Congr. Genet.* 1, 81-91.

Robichaud, C.S., Wong, J. and Sussex, I.M. (1980) Control of in vitro growth of viviporous embryo mutants of maize by abscisic acid. *Dev. Genet.* 1, 325-330.

Roeder, G.C. and Fink, G.R. (1983) Transposable elements in yeast. In *Mobile Genetic Elements.* ed. J.A. Shapiro, pp. 299-328. Academic Press, New York.

Roeder, G.S., Rose, A.B. and Perlman, R.F. (1985) Transposable element sequences involved in the enhancement of yeast gene expression. *Proc. Nat. Acad. Sci. USA* 82, 5428-5432.

Rogers, J.H. (1985) The origin and evolution of retroposons. *Int. Rev. Cytol.* 93, 187-279.

Rogers, J. and Wall, R. (1984) Immunoglobulin RNA rearrangements in B lymphocyte differentiation. *Adv. Immunol.* 35, 39-56.

Rogers, S.G., Brand, L.A., Holden, S.B., Sharps, E.S. and Brackin, M.I. (1983) Amplification of the aro A gene from *Escherichia coli* results in tolerance to the herbicide glyphosate. *Appl. Environ. Microbiol.* 46, 37-43.

Romer, A.S. (1958) Phylogeny and behaviour with special reference to vertebrate evolution. In *Behaviour and Evolution* ed. A. Roe, and G.F. Simpson. pp. 48-75. Yale University Press, New Haven.

Romer, A.S. (1966) *Vertebrate Paleontology,* 3rd Ed. University Press, Chicago.

Romer, A.S. (1968) *The Procession of Life.* World Publishing Company, Cleveland.

Rose, M.R. and Doolittle, W.F. (1983) Molecular biological mechanisms of speciation. *Science Wash.* 220, 157-162.

Rosenberg, E., Keller, K. and Dworkin, M. (1977) Cell density-dependent growth of *Myxococcus xanthus* on casein. *J. Bacteriol.* 129, 770-777.

Rosenberg, U.B., Schröder, C., Preiss, A., Kienlin, A., Coti, S., Riede, E. and Jäckle, H. (1986) Structural homology of the product of the *Drosophila* Krüppel gene with *Xenopus* transcription factor IIIA. *Nature (London)* 319: 336-339.

Rosenfeld, M.G., Mermod, J., Amara, S.G., Swanson, L.W., Sawchenko, P.E., Rivier, J., Vale, W.W. and Evans, R.M. (1983) Production of a novel neuropeptide encoded by the calcitonin gene via tissue-specific RNA processing. *Nature (London)* 304, 129-135.

Ross, H.H. (1966) *Understanding Evolution.* Prentice-Hall, New Jersey.

Rothschild, N.M.V., (1965) *A Classification of Living Animals.* Longmans, London.

Rubin, G.M. (1983) Dispersed repetitive DNAs in *Drosophila.* In *Mobile Genetic Elements* ed. J.A. Shapiro, pp. 329-361. Academic Press, New York.

Runnegar, B. (1982) The molecular clock date for the origin of the animal phyla. *Lethaia* 15, 199-205.

Sadoglu, P. (1967) The selective value of eye and pigment loss in Mexican cavefish. *Evolution* 21, 541-549.

Sandon, H. (1968) *Essays in Protozoology.* Hutchinson, London.

Saville, D.B.O. (1968) Possible interrelations between fungal groups. In *The Fungi,* ed. C.H. Ainsworth and A.S. Sussman, Vol. 3 pp. 659-675. Academic Press, New York.

Schindewolf, O.H. (1950) *Die Zeitfaktor in Geologie and Paläontologie.* Schweizerbart, Stuttgart.

Schindewolf, O.H. (1950) *Grundfragen der Paläontologie.* Schweizerbart, Stuttgart.

Schmidt, M., Glogger, K., Wirth, T., and Horsk, I. (1984) Evidence that a major class of mouse long terminal repeats (LTRS) resulted from recombination between exogenous retroviral LTRS and similar LTR-like elements (LTR-IS). *Proc. Nat. Acad. Sci. USA* 81, 6696-6700.

Schmidtke, J., Brennecke, H., Schmid M., Neitzel, H. and Sterling, K. (1981) Evolution of Muntzac DNA. *Chromosoma* 84, 187-193.

Schnepf, E. (1964) Zur Feinstruktur von *Geosiphon pyriforme.* Ein Versuch zur Deutung cytoplasmatischer Membrane Kompartmente. *Arch. Mikrobiol.* 49, 112-131.

Schönmuth, G. (1980) Genetic bases of animal breeding. *Proc. XIV Int. Congr. Genet.* 1, 152-160.

Schopf, J.W. and Oehler, D.Z. (1976) How old are the eukaryotes? *Science Wash.* 193, 47-49.

Selander, R.K. and Levin, B.R. (1980) Genetic diversity and structure in *Escherichia coli* populations. *Science (Wash.)* 210, 545-547.

Sepkoski, J.J. (1982) A compedium of fossil marine families. *Milwaukee Public Mus. Contrib.* No. 51, pp. 1-125.

Seyfert, H.M. and Willis, J.H. (1981) Molecular polymorphism of ciliary proteins from different species of the ciliate *Tetrahymena. Biochem. Genet.* 19: 385-396.

Shepherd, J.C.W., McGinnis, W., Carrasco, A.E., De Robertis, E.M. and Gehring, W.J. (1984) Fly and frog homoeo domains show homologies with yeast mating type regulatory proteins. *Nature (London)* 310, 70-71.

Sheridan, A.K. (1981) Crossbreeding and heterosis. *Anim. Breeding Abst.* 49, 131-144.

Shih, M.-C., Lazar, G. and Goodman, H.M. (1986) Evidence in favor of the symbiotic origin of chloroplasts: primary structure and evolution of tobacco glyceraldehyde-3-phosphate dehydrogenases. *Cell* 47, 73-80.

Shimkets, L. and Dworkin, M. (1981) Excreted adenosine is a cell density signal for the initiation of fruiting body formation in *Myxococcus xanthus Dev. Biol.* 84, 51-60.

Shorrocks, B. (1978) *The Genetics of Diversity.* Hodder and Stoughton, London.

Siegfried, P. (1954) Die Fisch-Fauna des westfalischen Ober-Senons. *Palaeontographica (A).* 106, 1-36.

Simchen, G., Winston, F., Styles, C.A. and Fink, G.R. (1984) TY-mediated gene expression of the Lys 2 and His 4 genes of *Saccharomyces cerevisiae* is controlled by the same SPT genes. *Proc. Nat. Acad. Sci. USA* 81, 2431-2434.

Simpson, G.G. (1964) *The Major Features of Evolution.* Columbia University Press, New York.

Simpson, G.G. (1964) *This View of Life.* Harcourt, Brace and World, New York.

Simpson, G.M., Durley, R.C., Kannangara, T. and Stout, D.G. (1978) The problem of plant breeders. In *Plant Regulation and World Agriculture.* ed. T.K. Scott. pp. 111-128. Plenum Press, New York.

Singer, S.R. and McDaniel, C.N. (1985) Selection of glyphosate-tolerant tobacco calli and the expression of this tolerance in regenerated plants. *Plant Physiol.* 78, 411-416.

Slightom, J.L., Blechi, A.E. and Smithies, O. (1980) Human fetal G γ- and A γ-globin genes: complete nucleotide sequences suggest that DNA can be exchanged between these duplicated genes. *Cell* 21, 627-638.

Smith, C. and Brampton P.R. (1977) Inheritance of reaction of halothane-induced anaesthesia in pigs. *Genet. Res.* 29, 287-292.

Smith, G.M. (1938) *Cryptogamic Botany* Vol. 1. *Algae and Fungi*. McGraw-Hill, New York.

Smith, G.R. (1983) Chi hotspots of generalized recombination. *Cell* 34, 709-710.

Sogin, M.L., Elwood, H.E. and Gunderson, J.H. (1986) Evolutionary diversity of eukaryotic small subunit rRNA genes. *Proc. Nat. Acad. Sci USA* 83, 1383-1387.

Sparrow, A.H. and Nauman, A.F. (1976) Evolution of genome size by DNA doublings. *Science* 192, 524-529.

Sperry R.W. (1958) Developmental basis of behavior. In *Behavior and Evolution.* ed. A. Roe and G.G. Simpson. pp. 128-139. Yale University Press, New Haven.

Spillett, J.J., Bunch, T.D. and Foote, W.C. (1975) The use of wild and domestic animals and the development of new genotypes. *J. Anim. Sci.* 40, 1009-1015.

Sprague, G.F. (1983) Heterosis in maize: Theory and practice. In *Heterosis: Reappraisal of Theory and Practice.* ed. R. Frankel, pp. 47–70. Springer, Berlin.

Spratt, B.G. (1988) Hybrid penicillin-binding proteins in penicillin-resistant strains of *Neisseria gonorrhoeae*. *Nature (London)* 332: 173–176.

Stack, S.M. and Brown, W.V. (1969) Somatic pairing, reduction and recombination: an evolutionary hypothesis of meiosis. *Nature (London).* 1275–1276.

Stahl, B.J. (1974) *Vertebrate History: Problems in Evolution.* McGraw-Hill, New York.

Stalker, D.M., Hiatt, W.R. and Comai, L. (1985) A single amino acid substitution in the enzyme 5-enolpyruvylshikimate-3-phosphate synthase confers resistance to the herbicide glyphosate. *J. Biol. Chem. J. Biol. Chem.* 260, 4724–4728.

Stanier, R.Y., Doudoroff, M. and Adelberg, E.A. (1972) *General Microbiology.* Macmillan, London.

Stanier, R.Y. and Van Niel, C.B. (1941) The main outlines of bacterial classification. *J. Bacteriol.* 42, 437–466.

Stanley, S.M. (1976) Fossil data and the Precambrian-Cambrian evolutionary transition. *Am. J. Sci.* 276, 56–76.

Stebbins, G.L. (1971) *Chromosomal Evolution in Higher Plants.* Arnold, London.

Stebbins, G.L. (1974) *Flowering Plants: Evolution above the Species Level.* Harvard University Press, Massachusetts.

Stebbins, G.L. (1982) Darwin to DNA, Molecules to Humanity. Freeman, San Francisco.

Stebbins, G.L. (1986) Gene action and morphogenesis in plants. In *Genetics, Development and Evolution.* ed. J.P. Gustafson, G.L. Stebbins and F.J. Ayala. pp. 29–48. Plenum Press, New York.

Stebbins, G.L. and Ayala, F.J. (1981) Is a new evolutionary synthesis necessary? *Science Wash.* 213, 967–971.

Stent, G.S. (1985) Thinking in one dimension: the impact of molecular biology on development. *Cell* 40, 1–2.

Sternberg, P.W. and Horvitz, H.R. (1982) Postembryonic nongonadal cell lineages of the nematode *Panagrellus redivivus*: description and comparison with those of *Caenorhabditis elegans*. *Dev. Biol.* 93, 181–205.

Sternberg, P.W. and Horvitz, H. R. (1982) The genetic control of cell lineage during nematode development. *Annu. Rev. Genet.* 18, 489–524.

Stevenson, T.M. and White, W.J. (1940) Investigations concerning the coumarin content of sweet clover. *Sci. Agr.* 21, 18–28.

Stoddart, J.L. (1966) Studies on the relationship between gibberellin metabolism and daylength in normal and non-flowering red clover (*Trifolium pratense L.*) *J. Exp. Bot.* 7, 96–107.

Stone, E.M. Rothblum, K.N. and Schwartz, R.J. (1985) Intron-dependent evolution of chicken glyceraldehyde phosphate dehydrogenase gene. *Nature (London) 313*, 498-500.

Storchová, H., Meyer, J. and Doskocil, J. (1985) An electron microscopic heteroduplex study of the sequence relations between the bacteriophages LP 52 and theta. *Mol. Gen. Genet.* 199, 476-480.

Strong, W.J. (1921) Greenhouse cucumber breeding. *Proc. Am. Soc. Hort. Sci.* 18, 271-273.

Stutz, E. (1984) The chloroplast genome of algae. *Bot. Helvetia* 94, 145-159.

Südhof, T.C., Goldstein, J.L., Brown, M.S. and Russell, D.W. (1985) The LDL receptor genes: a mosaic of exons shared with different proteins. *Science Wash.* 228, 815-822.

Sulston, J. and Horvitz, H.R. (1981) Abnormal cell lineages in mutants of the nematode *Caenorhabditis elegans*. *Dev. Biol.* 82, 41-55.

Swain, T. (1974) Biochemical evolution in plants. *Compre. Biochem.* 29A, 125-302.

Teeri, T.T. Lehtovaara, P., Kauppinen, S., Salovuori, I. and Knowles, J. (1987) Homologous domains in *Trichoderma reesei* cellulolytic enzymes: gene sequence and expression of cellobiohydrolase II. *Gene* 51, 43-52.

Temin, H.M. (1985) Reverse transcription in the eukaryotic genome: retroviruses, pararetroviruses, retrotransposons, and retrotranscripts. *Mol. Biol. Evol.* 2, 455-468.

Temin, H.M. and Mizutani, S. (1970) RNA-dependent DNA polymerase in virions of rous sarcoma virus. *Nature (London)* 226, 1211-1213.

Terrière, L.C. and Yu, S.J. (1973) Insect juvenile hormones: induction of detoxifying enzymes in the house fly and the detoxification by house fly enzymes. *Pestic. Biochem. Physiol.* 3, 96-107.

Thiesen, M., Stief, A. and Sippel, A.E. (1986) The lysozyme enhancer: cell-specific activation of the chicken lysozyme gene by a far-upstream DNA element. *EMBO J.* 5, 719-724.

Thomas, G.H., Siegfried, E. and Elgin, S.C.R. (1985) DNase I hypersensitive sites: a structural feature of chromatin association with gene expression. In *Chromosomal Proteins and Gene Expression.* ed. G. Reek, G. Goodwin and P. Puigdomenech. Plenum Press, New York.

Thompson, C.J. and Gray, G.S. (1983) Nucleotide sequence of a streptomycete aminoglycoside phosphotransferase gene and its relationship to phosphotransferases encoded by resistance plasmids. *Proc. Nat. Acad. Sci. USA* 80, 5190-5194.

Thompson, D'Arcy, W. (1942) *On Growth and Form.* Cambridge University Press, Cambridge.

Thornback, J. and Jenkins, M. (Compilers) (1982) *The I.U.C.N. Mammal Red Data Book.* I.U.C.N., Gland, Switzerland.

Thuriaux, P. (1985) Direct selection of mutants influencing gene conversion in the yeast *Saccharomyces pombe*. *Mol. Gen. Genet.* 199, 365-371.

Tiffney, B.H. and Barghoorn, E.S. (1974) The fossil record of the fungi. *Occas. Pap. Farlow Herb. Cryptogam. Bot.* No. 7 pp. 1-42.

Tinbergen, N. (1960) Behaviour, systematics and natural selection. In *Evolution after Darwin* ed. S. Tax. Vol. 1. pp. 595-613. University of Chicago Press, Chicago.

Tingle, M., Singh, A.J., Henry, S.A. and Halvorson, H.O. (1973) Ascospore formation in yeast. *Symp. Soc. Gen. Microbiol.* 23, 209-245.

Tobin, E.M. and Silverthorne, J. (1985) Light regulation of gene expression in higher plants. *Annu. Rev. Plant Physiol.* 36, 569-593.

Todaro, G.J., Callahan, R., Rappe, U.R. and De Larco, J.E. (1980) Genetic transmission of retroviral genes and cellular oncogenes. *Proc. R. Soc. London* B 291, 367-385.

Towe, K.M. (1970) Oxygen-collagen priority and the early metazoan fossil record. *Proc. Nat. Acad. Sci. USA* 65, 781-788.

Truffa-Bachi, P., Guiso, N., Cohen, G.N., Theze, J. and Burr, B. (1975) Evolution of biosynthetic pathways: immunological approach. *Proc. Nat. Acad. Sci. USA* 72, 1268-1271.

Tschudi, C., Young, A.S., Ruben, K., Patton, C.L. and Richards, F.F. (1985) Calmodulin genes in trypanosomes are tandemly repeated and produce multiple mRNAs with a common 5' leader sequence. *Proc. Nat. Acad. Sci. USA* 82, 3998-4002.

Tso, J.Y., Van den Berg, D.J. and Korn, L.J. (1986) Structure of the gene for *Xenopus* transcription factor TFIIIA. *Nuc. Acids Res.* 14, 2187-2200.

Twitty, V.C. (1949) Developmental analysis of amphibian pigmentation. *Growth Symp.* 9, 133-161.

Ullrich, R.C. and Raper, J.R. (1977) Evolution of genetic mechanisms in fungi. *Taxon* 26, 169-179.

Ulrych, T.J. (1967) Oceanic basalt leads: a new interpretation and an independent age for the earth. *Science* 158, 252-256.

Valentine, J.W. (1973) *Evolutionary Paleoecology of the Marine Biosphere.* Prentice-Hall, New Jersey.

Vandel, A. (1972) La répartition des Oniscoides (crustacés, isopods terrestres) et la dérive des continents. *C.R. Acad. Sci.* D 275, 2069-2072.

Van Etten, H.D., Matthews, D.E. and Smith, D.A. (1982) Metabolism of phytoalexins. In *Phytoalexins* ed. J.A. Bailey and J.W. Mansfield. pp. 181-217, Blackie, Glasgow.

Van Tuyl, J.M. (1977) Genetics of fungal resistance to systemic fungicides. *Meded. Landbouwhogesch. Wageningen Ser.* 77-2.

Van Valen, L. (1973) A new evolutionary law. *Evol. Theory* 1, 1-30.

Van Valen, L. (1973) Are categories in different phyla comparable? *Taxon* 22, 333-373.

Varmus, H.E. (1983) Retroviruses. In *Mobile Genetic Elements.* ed. J.A. Shapiro. pp. 411-503. Academic Press, New York.

Vavilov, N.L. (1951) The origin, variation, immunity and breeding of cultivated plants (K.S. Chester, transl.) *Chron. Bot.* 13, 1-366.

Vidal, G. and Knoll, A.H. (1983) Proterozoic plankton. *Mem. Geol. Soc. Am.* 161, 265-277.

Vilgalys, R.J. and Johnson, J.L. (1987) Extensive genetic divergence associated with speciation in filamentous fungi. *Proc. Nat. Acad. Sci. USA* <u>84</u>, 2355-2358.

Volkenstein, M.V. (1965) Coding of polar and non-polar amino-acids. *Nature (London)* <u>207</u>, 294-295.

Voronin, L.G. (1962) Some results of comparative physiological investigations of higher nervous activity. *Psychol. Bull.* <u>59</u>, 161-195.

Waddington, C.H. (1953) Genetic assimilation of an acquired character. *Evolution* <u>7</u>, 118-126.

Wagner, M. (1986) A consideration of the origin of processed pseudogenes. *Trends in Genet.* May 134-137.

Wakimoto, B.T., Kalfayan, L.J. and Spradling, A.C. (1986) Developmentally regulated expression of *Drosophila* chorion genes introduced at diverse chromosomal positions. *J. Mol. Biol.* <u>187</u>, 33-45.

Walker, J.C.G., Klein, K., Schidlowski, M., Schopf, J.W., Stevenson, D.J. and Walter, M.R. (1983) Environmental evolution of the archaen-early proterozoic earth. In *Earth's Earliest Biosphere.* ed. J.W. Schopf pp. 220-290. Princeton University Press, New Jersey.

Wallace, D.G., Maxon, L.R. and Wilson, A.C. (1971) Albumin evolution in frogs: a test of the evolutionary clock hypothesis. *Proc. Nat. Acad. Sci. USA* <u>68</u>, 3127-3129.

Walker, T.R. (1969) The evolution of the *Argopecten gibbus* stock (Mollusca: Bivalvia), with emphasis on the Tertiary and Quaternary species of eastern North America. *J. Paleont.* <u>43</u>, (suppl.) 1-125.

Wallis, M. (1975) The molecular evolution of pituitary hormones. *Biol. Rev.* <u>50</u>, 39-98.

Walter, M.R. (1987) The timing of major evolutionary innovations from the origin of life to the origins of the Metaphyta and Metazoa: The geological evidence. In *Rates of Evolution.* ed. K.W.S. Campbell and M.F. Day. pp. 15-38. Allen and Unwin, London.

Walter, M.R., Oehler, J.H. and Oehler, D.Z. (1976) Megascopic algae 1300 million years old from the Belt supergroup, Montana: a reinterpretation of Walcott's Helminthoidichnites. *J. Paleont.* <u>50</u>, 872-881.

Wangsness, P.J., Martin, R.J. and Gahagan, J.H. (1977) Insulin and growth hormone in lean and obese pigs. *Am. J. Physiol.* <u>233</u>, E 104.

Wardlaw, C.W. (1965) *Organization and Evolution in Plants.* Longmans, London.

Wareing, P.F. (1978) Growth regulators and assimilate partitioning. In *Plant Regulation and World Agriculture.* ed. T.K. Scott. pp. 309-317. Plenum Press, New York.

Warren, J.M. (1957) The phylogeny of maze learning: (1) Theoretical orientation. *Brit. J. Anim. Behav.* <u>5</u>, 90-93.

Watanabe, T. (1963) Infective heredity of multiple drug resistance in bacteria. *Bacteriol. Rev.* <u>27</u>, 87-115.

Watkins, A.E. and Ellerton, S. (1940) Variation and genetics of the awn in *Triticum.* *J. Genet.* <u>40</u>, 243-270.

Webb, S.D. (1969) Extinction-origination equilibria in the late Cenozoic land mammals of North America. *Evolution* 23, 688-702.

Weber, F., de Villiers, J. and Schaffner, W. (1984) An SV 40 'enhancer trap' incorporates exogenous enhancers or generates enhancers from its own sequences. *Cell* 36, 983-992.

Webster, R.G., Laver, W.G., Air, G.M. and Schild, G.C. (1982) Molecular mechanisms of variation in influenza virus. *Nature (London)* 296, 115-121.

Weisblat, D.A. (1985) Segmentation and commitment in the leech embryo. *Cell* 42, 701-702.

Weitzman, P.D.J. and Dunmore, P. (1969) Regulation of citrate synthase activity by α-ketoglutarate. Metabolic and taxonomic significance. *F.E.B.S. Lett* 3, 265-267.

Wessler, S.R. and Varagona, M.J. (1985) Molecular basis of mutations at the waxy locus of maize: correlation with the fine structure genetic map. *Proc. Nat. Acad. Sci. USA* 82, 4177-4181.

White, J.G., Southgate, E., Thomson, J.N. and Brenner, S. (1983) Factors that determine connectivity in the nervous system of *Caenorhabditis elegans*. *Cold Spring Harbor Symp. Quant. Biol.* 48, 633-640.

White, M.J.D. (1973) *Animal Cytology and Evolution*, 3rd ed. Cambridge University Press, Cambridge.

Whittaker, R.H. (1969) New concepts of kingdoms of organisms. *Science* 163, 150-160.

Wilcken-Bergmann, B. von and Müller-Hill, B. (1982) Sequence of gal R gene indicates a common evolutionary origin of lac and gal repressor in *Escherichia coli*. *Proc. Nat. Acad. Sci. USA* 79, 2427-2431.

Wildeman, A.G. (1988) A putative ancestral actin gene present in a thermophilic eukaryote: novel combination of intron positions. *Nuc. Acids Res.* 16.

Williamson, V.M., Young, E.T. and Ciriacy, M. (1981) Transposable elements associated with constitutive expression of yeast alcohol dehydrogenase II. *Cell* 23, 605-614.

Willis, J.C. (1940) *The Course of Evolution by Differentiation or Divergent Mutation rather than by Selection*. Cambridge University Press, Cambridge.

Wilson, A.C. (1976) Gene regulation in evolution. In *Biochemical Evolution*. ed. F.J. Ayala. pp. 225-234. Sinauer, Massachusetts.

Wilson, A.C., Bush, G.L., Case, S.M. and King, M.C. (1975) Social structuring of mammalian populations and rate of chromosome evolution. *Proc. Nat. Acad. Sci. USA* 72, 5061-5065.

Wilson, E.O. (1971) *The Insect Societies*. Bleknap Press, Massachusetts.

Wilson, E.O. (1975) *Sociobiology. The New Synthesis*. Bleknap Press, Massachusetts.

Wilson, P.W., Rogers, J., Harding, M., Pohl, V., Pattyn, G. and Lauson, D.E.M. (1988) Structure of the chick chromosomal genes for calbindin and cabretinin. *J. Mol. Biol.* 200, 615-625.

Winston, F., Chaleff, D.T., Valent, B. and Fink, G.R. (1984) Mutations affecting Ty-mediated expression of the HIS 4 gene of *Saccharomyces cerevisiae*. *Genetics* 107, 179-197.

Winteringham, F.P.W. (1957) Comparative biochemical aspects of insecticidal action. *Chem. and Ind. (London)* pp. 1195-1202.

Wolpert, L. (1970) Developing cells know their place. *New Sci.* 14 May, pp. 322-325.

Wolter, F.P., Fritz, C.C., Willmitzer, L., Schell, J. and Schreier, P.H. (1988) rbsS genes in *Solanum tuberosum*: conservation of transit peptide and exon shuffling during evolution. *Proc. Nat. Acad. Sci. USA* 85, 846-850.

Womack, J.E. and Moll, Y.D. (1986) Gene map of the cow: conservation of linkage with mouse and man. *J. Hered.* 77, 2-7.

Wu, C.I. and Li, W.-H. (1985) Evidence for higher rates of nucleotide substitution in rodents than in man. *Proc. Nat. Acad. Sci. USA* 82, 1741-1745.

Wu, J.B., Kingston, R.E. and Morimoto, R.J. (1986) Human HSP70 promoter contains at least two distinct regulatory domains. *Proc. Nat. Acad. Sci. USA* 83, 629-633.

Wu, T.T., Lin, E.C.C. and Tanaka, S. (1968) Mutants of *Aerobacter aerogenes* capable of utilizing xylitol as a novel carbon. *J. Bacteriol.* 96, 447-456.

Wyles, J.S., Kunkel, J.G. and Wilson, A.C. (1983) Birds, behavior, and anatomical evolution. *Proc. Nat. Acad. Sci. USA* 80, 4394-4397.

Yadav, N., McDevitt, R.E., Benard, S. and Falco, S.C. (1986) Single amino acid substitutions in the enzyme acetolactate synthase confer resistance to the herbicide sulfometuron methyl. *Proc. Nat. Acad. Sci. USA* 83, 4418-4422.

Yagi, Y. and Clewell, D.B. (1976) Plasmid-determined tetracycline resistance in *Streptococcus faecalis*: tandemly repeated resistance determinants in amplified forms of p AM α 1 DNA. *J. Mol. Biol.* 102, 583-600.

Yaguchi, M., Roy, C., Rollin, C.F., Paice, M.G. and Jurasek, L. (1983) A fungal cellulase shows sequence homology with the active site of hen egg-white lysozyme. *Biochem. Biophys. Res. Comm.* 116, 408-411.

Yamamoto, K.R. (1985) Steroid receptor regulated transcription of specific genes and gene networks. *Annu. Rev. Genet.* 19, 209-252.

Yamamoto, T., Gojobori, T., and Yokota, T. (1987) Evolutionary origin of pathogenic determinants in enterotoxigenic *Escherichia coli* and *Vibrio cholerae* 01. *J. Bacteriol.* 169, 1352-1357.

Yeh, W.K. and Ornston, L.N. (1980) Origins of metabolic diversity: substitution of homologous sequences into genes for enzymes with different catalytic activities. *Proc. Nat. Acad. Sci. USA* 77, 5365-5369.

Young, J.Z. (1950) *The Life of Vertebrates*. Clarendon Press, Oxford.

Young, M.W. (1979) Middle repetitive DNA: a fluid component of the *Drosophila* genome. *Proc. Nat. Acad. Sci. USA* 76, 6274-6278.

Young, R.A., Hagenbuchle, O. and Schibler, U. (1981)   A single mouse α-amylase gene specifies two different tissue-specific mRNAs.   *Cell* 23, 451-458.

Yu, S.J. (1982) Host plant induction of glutathione-S-transferase in the fall armyworm.   *Pestic. Biochem. Physiol.* 18, 101-106.

Yuan, K., Johnson, W.C., Tepper, D.J. and Setlow, P. (1981) Comparisons of various properties of low molecular weight proteins from dormant spores of several *Bacillus* species.   *J. Bacteriol.* 46, 965-971.

Zackson, S.L. (1984)   Cell lineage, cell-cell interaction, and segment formation in the ectoderm of a glossiphoniid leech embryo.   *Dev. Biol.* 104, 143-160.

Zeevaart, J.A.D. (1978)   Phytohormones and flower formation.   In *Phytohormones and Related Compounds - a Comprehensive Treatise* ed. D.S. Letham, R.O. Goodwin and T.J.V. Higgins, Vol. II pp. 291-328. Elsevier, Amsterdam.

Zeuner, F.E. (1963)   *A History of Domesticated Animals.*   Hutchinson, London.

Ziaie, Z. and Suyama, Y. (1987) The cytochrome oxidase subunit I gene of *Tetrahymena*:   a 57 amino acid NH_2-terminal extension and a 108 amino acid insert.   *Curr. Genet.* 12: 357-368.

Zieg, J., Silverman, M., Hilman, M. and Simon, M. (1977) Recombinational switch for gene expression.   *Science Wash.* 196, 170-172.

Ziff, E.B. (1985) Splicing in adenovirus and other animal viruses.   *Int. Rev. Cytol.* 93, 327-358.

Zobel, R.W. (1973) Some physiological characteristics of the ethylene requiring mutant diageotropica.   *Plant Physiol.* 52, 385-389.

Zohary, D. (1969) The progenitors of wheat and barley in relation to domestication and agricultural dispersal in the Old World.   In *The Domestication and Exploitation of Plants and Animals.* ed. P.J. Ucko and G.W. Dimbleby. pp. 47-66. Duckworth, London.

Zuckerkandl, E. (1978) Multilocus enzymes, gene regulation, and genetic sufficiency.   *J. Mol. Evol.* 12, 57-89.

Zuckerkandl, E. and Pauling, L. (1965)   Evolutionary divergence and convergence in proteins. In *Evolving Genes and Proteins.* ed. V. Bryson and H.J. Vogel. pp. 97-166. Academic Press, New York.

INDEX